CLIMATE FOR THE LAYMAN

Anthony Bright-Paul

Climate For The Layman

Anthony Bright-Paul

An Authors OnLine Book

Copyright © Anthony Bright-Paul 2014

Cover picture by kind permission of
Principia Scientific International

All rights reserved. No part of this publication may be reproduced, stored in a retrieval system, or transmitted in any form or by any means, electronic, mechanical, photocopy, recording or otherwise, without prior written permission of the copyright owner. Nor can it be circulated in any form of binding or cover other than that in which it is published and without similar condition including this condition being imposed on a subsequent purchaser.

Published by: Authors
OnLine Ltd 19 The
Cinques Gamlingay,
Sandy Bedfordshire
SG19 3NU England

www.authorsonline.co.uk

Table of Contents

	Date	Title	Author(s)
1	Preface	Preface	Anthony Bright-Paul
2	07.04.2006	There is a Problem	Professor Bob Carter
3	06.04.2007	Tim Ball Letter 1	Professor Tim Ball
4	08.04.2007	Tim Ball Letter 2	Professor Tim Ball
5	08.07.2007	Tim Ball Letter 3	Professor Tim Ball
6	09.11.2007	My Nobel Moment	Professor John Christy
7	16.12.2007	Tim Ball Letter 4	Professor Tim Ball
8	17.12 2007	Causes of Climate Change	Anthony Bright-Paul
9	04.02.2008	Tim Ball Letter 5	Professor Tim Ball
10	23.03.2008	The Sea is a Fizzy Drink	Anthony Bright-Paul
11	03.05.2008	Scientific Consensus	Anthony Bright-Paul
12	04.06.2008	The Death Blow to AGW	Stephen Wilde
13	18.07.2008	No Smoking Hotspot	Dr David Evans
14	05.02.2009	Black Magic	Anthony Bright-Paul
15	22.07.2009	Models of Illusion	John Droz Jr
16	20.08.2009	Singing from the same Hymn Sheet	Anthony Bright-Paul
17	21.02.2009	Emails James Peden	James Peden
18	04.09.2009	Two Jags	Richard Littlejohn
19	26.09.2009	Climate Change, what to we know?	Anthony Bright-Paul
20	21.10.2009	Apocalypse now in 50 days	Anthony Bright-Paul
21	26.11.2009	Climate Fears	Piers Corbyn
22	21.12.2009	All Change	Johnny Ball
23	14.12.2009	For whom the bell tolls	Anthony Bright-Paul
24	09.02.2010	What is Global Warming?	Anthony Bright-Paul
25	30.03.2010	Global warming and that 2nd Law	Anthony Bright-Paul
26	16.11.2010	Weasel Words	Anthony Bright-Paul
27	22.02.2011	Beware the Global Fascists	Johnny Ball
28	11.03.2011	My Lack of Physics	ABP + Hans Schreuder
29	14.03.2011	Global Warming: 10 Little facts	Professor Bob Carter
30	20.09.2011	Standard Atmosphere	ABP + Hans Schreuder
31	26.09.2011	Two Halves make one whole	Anthony Bright-Paul
32	02.12.2011	Simple Arithmetic	Anthony Bright-Paul
33	28.03.2011	The Laws of Physics	Anthony Bright-Paul
34	10.05.2011	Greenhouse Gasses Cool ….	Hans Schreuder + Assocs
35	31.07.2011	The Underwhelming Evidence	Anthony Bright-Paul
36	16.09.2011	Inverse Square Law, General	John Etherington
37	05.11.2011	Get your heads round this	Hans Schreuder + ABP
38	19.02.2012	If they were serious	Anthony Bright-Paul
39	17.04.2012	Heat Generation - Revised	Anthony Bright-Paul

40	29.04.2012	7 Billion Machines	Anthony Bright-Paul
41	26.07.2013	History of wrong conclusions	Anthony Bright-Paul
42	27.08.2013	Living on the Moon	Anthony Bright-Paul
43	19.10.2013	Shall I tell you a secret?	Anthony Bright-Paul
44	30.10.2013	What's in the space?	ABP + Philip Foster
46	01.11.2103	You can't heat nothing!	Anthony Bright-Paul
47	10.02.2014	Can anyone doubt?	Anthony Bright-Paul
48	20.02.2014	The Fundamental Error	Anthony Bright-Paul
49	26.02.2014	Does Wales need a single Turbine?	Anthony Bright-Paul
50	27.02.2014	Wind Energy	John Droz Jr
51	17.03.2014	Can Air heat itself?	Anthony Bright-Paul
52	22.03.2014	Arse about Face	Anthony Bright-Paul
53	01.03.2014	Why are they lying?	Anthony Bright-Paul
54	01.04.2014	Global Warming is Rubbish	Professor Les Woodcock
55	27.05.2014	The Trouble with Climate Change	Lord Nigel Lawson
56	04.05.2014	Everywhere and All-at-once	Anthony Bright-Paul
57	25.02.2014	If you want crap science	Anthony Bright-Paul
58	10.05.2014	Night and Day	Anthony Bright-Paul
59	17.05.2014	Solar power Distribution	Anthony Bright-Paul
60	12.07.2014	Heat and a Thermos Flask	Anthony Bright-Paul
61	09.08.2014	The Sand Lizard	Anthony Bright-Paul
62	31.07.2014	Conclusion: It all depends…	Anthony Bright-Paul

Preface

This book 'Climate for the Layman' is written by a layman for laymen. It disputes entirely the idea that climate is a subject only for scientists. On the contrary this book is about logic and reason, which a normal unprejudiced and non-hysterical person can understand. Of course, basic principles of science are necessary, but it must be science combined with logic and a healthy dose of scepticism.

It is written for those who are naturally sceptical – that is for those who question, 'What does this mean? Is that a fact? Is that correct?' For what is the opposite of sceptical? The opposite of sceptical is gullible. So this book is not written for those who are easily gulled, who will accept any old thing at face value.

Effectively this book began to be written on the 8th of March 2007 when the author first saw the documentary 'The Great Global Warming Swindle' on Channel 4. This documentary appealed to this author precisely because it appealed to reason, and this view has not changed. On the contrary it has been reinforced. I am entirely grateful to Martin Durkin for stimulating me to question the whole business of man-made Global Warming, or Anthropogenic Global Warming, which is often abbreviated to AGW.

Let us start with definitions. What does the word 'global' mean? Are we talking about the entire world with its four levels of atmosphere? If we are to include the four levels of atmosphere we then have to ask are they all warming? What is the standard? What is the benchmark? From what date must one start?

The Sun sends radiation through Outer Space some 94 million miles to reach the Earth. So is Outer Space tremendously hot? The answer is simple, though rarely understood. Outer Space is neither hot nor cold because it is empty, it is a vacuum. There has to be substance or mass to get hot or to be heated.

So the Sun does not send heat through space, since if it did, then Outer Space would be hot – it sends radiation and radiation has to encounter something, has to encounter mass to produce heat.

So let us descend through our atmosphere. The Thermosphere which is the outer layer is very near empty also, so it near vacuum. But the question is simple: is the Thermosphere warming? In order to declare that the Thermosphere is warming one first has to establish a time frame. Let us take 1900. Can we say what was the temperature of the Thermosphere in 1900? Clearly, by logic alone, it is and it was totally impossible to establish any such temperature then and furthermore to establish any such temperature for the Thermosphere now.

So we may ask likewise, is the Mesosphere warming? Is the Stratosphere warming? And finally is the Troposphere warming?

Just how does one measure warming? The only way that we know is by use of a thermometer. So where does one place the thermometer in order to find a temperature? Since the four levels of the atmosphere together are some 62 miles high, or 100 kilometres, these atmospheres have a huge range of temperatures. In the Troposphere alone, the level where our own weather occurs, which is perhaps 7 miles high, the temperatures are constantly changing. The hottest part of the atmosphere is at the base, at sea level, while the temperatures decline with altitude by 2° C for every 1,000 feet of altitude. So how does one measure the temperature of something that has a multiplicity of temperatures and is constantly on the move? It is clearly impossible.

Imagine a column of air immediately above one's head to the edge of the Troposphere. This column wherever one may start is getting colder and colder with altitude. How does one measure that? Or, should we measure along a line of latitude? What, the Equator, or the Arctic Circle? Or should we then measure along a line of longitude, from the North Pole through the Tropics and down to the South Pole? Would that be satisfactory, when say the temperature at the North Pole would effectively cancel out the temperature of say Accra?

So we resort to weather stations dotted about on the surface of our earth, which is uneven with hills and mountains and steep valleys, with open country, huge forests, deserts, ice-caps and huge metropolitan areas and with some 70% oceans.

It is true that such institutions as the Goddard Institute, or NASA and the Climate Research Unit in England issue graphs to show that the atmosphere is warming or cooling, as www.climate4you.com but these figures have strict limitations. There is no real attempt to measure the temperature of the atmosphere for the simple reason that no such measure is possible. Radiosonde balloons are sent aloft from many parts of the world, and these only confirm the huge variations in temperature as they ascend up into the Stratosphere transmitting information as they go. Aircraft of all sorts have monitors aboard by means of which passengers can observe the speed and also the changing temperatures outside.

So what do these graphs refer to? They are temperatures taken on the ground or rather 5 foot above the ground in special louvred boxes, called Stephenson Screens. These readings are all lumped together and an average is taken in a strange and arcane way, with the use of anomalies. So can we infer for one moment that these measurements are measuring the atmosphere of this globe, of this planet, upon the crust of which we are living? No such measurement is possible. The anomaly against which these measurements are made is just that – an anomaly. It is a sort of standard average at near ground level against which other averages are measured in order to ascertain whether the these averages are moving up or down. The imagination boggles!

Furthermore, where once this average was the average of some 6,000 stations, it has been reduced to some 1,500 in order that weather stations on hills and mountainsides are excluded. So logically these figures, these averages, bear very little resemblance to the realities we all experience every day.

As everybody knows, who follows the weather forecasts, in no matter what country, the temperatures are changing minute by minute, as the Earth rotates upon its axis in its journey round the Sun, and as night follows day and the seasons succeed one another.

If it were possible to take the temperature of the Earth we should be able to enquire, 'What was the Global mean temperature last Saturday? And how does it compare with the temperature today that is Monday, 7th July 2014?' No such figures are available. We can only see in hindsight when there have been prolonged periods when large portions of say the northern hemisphere have been covered by snow and ice, and other long periods when the Earth has been comparatively warm at the surface of the planet where we all exist.

In my dictionary 'Global' is defined as 'worldwide'. So let us ask ourselves the question - Has there been a worldwide warming of 0.07° Celsius? Has there been a uniform increase in temperatures worldwide? The answer is simple. It is utterly impossible to make any such declaration, for the aforesaid reasons. It is completely impossible to measure the temperature of the atmosphere which is 100 kilometres high and which has a huge range of temperatures in a continuous state of flux.

I see in my dictionary that there is also added a definition of Global Warming – **the increase in temperature of the Earth's atmosphere caused by the Greenhouse Effect.** Read this carefully. How can anyone who has a minimum of logic be taken in by such a definition? Let us leave aside for one moment the Greenhouse Effect, which will be dealt with later in this book. Firstly is there such a thing as the 'temperature of the atmosphere?' Clearly there is no such entity. For there to be an 'increase' in temperature of the atmosphere there would have to be a starting point. But there is no starting point, there is no norm, unless one is chosen arbitrarily, nor is there any means whatsoever of measuring the immeasurable. If at this very moment the temperature in Farnborough is 20°C at 5 foot above ground level, what is the temperature at 5,000 feet? It would be 10°C, and at 10,000 feet it would be zero degrees. As the sun is setting the temperature at ground level is declining, and as that temperature is declining so also the range of temperatures is also declining. Since there is no such thing as a temperature of the whole Earth all talk of Global Warming is simply illogical, ill thought out and needs to be discarded for the sake of clarity. The Globe is warming and cooling in different locations concurrently every minute of the day and night.

Since it is abundantly clear that there is no one temperature of the atmosphere all talk of Anthropogenic Global Warming is simply an exercise in fatuity. Let us go farther. Are the soils of the earth warming? Are the rocks warming? Are the oceans warming? Are the mountains warming? Are the volcanoes warming? Is the lava in the volcanoes hotter than whenever? Are the geysers in the USA and New Zealand hotter today than they were last week, or in 1900? Is the water that arises in vents from the ocean floor is that hotter? Need I go on? None of these questions can be answered, unless one had an instant and worldwide knowledge of everywhere and all at once.

Just as it is impossible to measure the temperature of the atmosphere or the Lower Atmosphere so also it is impossible to take the temperature of all the sands, and all the soils and all the rocks upon the surface of this blessed planet. How can one measure flux? How can one measure something that is constantly changing?

Sure scientists are trying to figure out the huge complexities of the climate systems, but for all the computer modelling it is for the most part just inspired guesswork. It is true that the University of Alabama and Remote Sensing Systems use satellites, but these only infer temperatures – no actual temperature readings are made. So although these satellites may show no or little warming, along come the Warmists to correct the data. Such manifest interference should not be tolerated.

With all the hysteria about Global Warming the question of what warms the atmospheres seems to be forgotten. Just how does the air get hot? Why is the air at 33,000 feet where most of our aeroplanes now fly, (as near as they can to the Tropopause, in order to be above the weather), why is it as low as minus 50 centigrade? How can it be so cold when it is nearer the Sun? Why should the atmosphere be warmer at sea level than it is high up? Why is there snow on the tops of mountains? Why is Mt Blanc covered with a cap of snow winter and summer? Why is Mt Everest not the hottest place on the planet, instead of one of the coldest?

In compiling this book, which is in fact a collection of essays and articles written over a period of some 7 years since I watched Martin Durkin's documentary I have had the good fortune to receive answers to my many questions from some of the finest brains on this planet. Somehow or other I got in touch with Professor Tim Ball in Canada, since I had seen him in the aforesaid documentary. I cannot remember how I tracked him down, but a series of his emails are included in this book. Professor Bob Carter of Queensland, Australia, for some while took it upon himself to encourage me in my writing, and visited with me in my home when he had occasion to come to England. He was an expert on precisely where does one start. If one started with the height of the Medieval Warm period, then we are cooling, but if we start at the end of the Little Ice Age then we are warming. But cooling or warming, what has that to do with man? I was also fortunate to attend a lecture in London by Professor Ian Plimer, having already read his book Heaven+Earth.

Subsequently I have met and corresponded with a number of scientists, the chief of whom has been Hans Schreuder, who has painstakingly answered my questions, while vetting my manuscripts. His explanations form the very core of this book. I am indebted also to Philip Foster, Piers Corbyn, Dr Philip Bratby, Astrophysicist Jim Peden, Geophysicist Norm Kalmanovitch, Alan Siddons, John Droz Jr, Gary Novak, Johnny Ball the mathematician, Lord Lawson, Meteorologist Stephen Wilde and many others. However, has their science differed substantially from the science of the Alarmists? Has their knowledge been equal among themselves?

The principal differences among them are not on matters of fact, but on the application of logic. So the conclusions arrived at are very often different and subject to contentious argument.

Let us take a simple example. If I bring some water in a kettle to the boil at 100°C and make a pot of tea, the water in the pot will inevitably cool, according to the 2nd law of thermodynamics. However if, as is the custom with many older people, I put a woollen cosy over the teapot the rate of cooling will decrease slightly. Will the woollen cosy increase the temperature of the water in the teapot? Of course not! If I were to apply two cosies instead of just one, would that increase the temperature one jot? Clearly not a whit. The cosy does not generate heat. In the same way Water Vapour in the atmosphere can reduce the rate of cooling. But can this greenhouse gas make anything hotter? Such an idea is fallacious – it is simply bad logic.

Let us not forget also that a cloud passing between us and the sun can also produce cooling, which is demonstrable to any child on a hot summer's day. If the gases of the atmosphere reduce the rate of cooling they also reduce the rate of warming. Insulation is a two-edged sword.

The same argument applies to Carbon Dioxide. It is argued that Carbon Dioxide traps the heat emanating from the earth's surface, that is, it inhibits heat loss. Not even the Alarmists would argue that Carbon Dioxide generates heat. So it does not make anything hotter, it may just slow the rate of cooling. Will the addition of a very slight increase in atmospheric Carbon Dioxide increase the temperature of the atmosphere, lower, higher or whatever? No more than can a double tea cosy. This is not a matter of science, but purely a matter of applied logic. And one must not forget for one moment that the atmosphere does not have one temperature, but a huge gradient.

In any case we must ask a simple question: Does the atmosphere warm the Earth, or does the Earth and the Oceans warm the atmosphere? Once we acknowledge the fact that the atmosphere cools with altitude, then we are bound by logic alone to concede that Earth warms atmosphere and it does that largely by conduction.

The Sun warms the Earth and the Earth warms the atmosphere.

Once we realise this sublime truth all the nonsense about Carbon Dioxide in the atmosphere causing any warming whatsoever is clearly shown to be absolutely pseudo-scientific nonsense. One hundred tea cosies will never make a pot of tea hotter. A slight increase in atmospheric Carbon Dioxide will not and cannot produce any warming, but can be hugely beneficial to a green planet. This is true greening, not the idiotic posturing of the so-called Greens.

It is sometimes argued even by respected Skeptics that 'climate sensitivity' refers to the amount by which the atmosphere will warm with an increase in Carbon Dioxide. But even the Alarmists do not propose that Carbon Dioxide produces heat – on the contrary they are at pains to point out that it absorbs and emits a lower infrared radiation from the Earth. So we have to ask, what is the causative factor? In each and every case the causative factor is the infrared radiation whether from Sun or from Earth.

Some eminent scientists with good degrees sometimes use the phrase 'hotter than it would otherwise be', as they refer to a supposed back radiation from the atmosphere. Even a layman with a minimum sense of logic can see how spurious such a statement is. If I turn on my oven to 250°C then that oven will get hotter than it was before heat was being generated. But can Carbon Dioxide, which is a trace part of the mixture of gases called air, generate heat? Not even the most besotted Alarmist would claim that.

Many matters of fact are agreed by both Alarmists and Skeptics. It is agreed for example that the two main constituents of the atmosphere are Nitrogen and Oxygen, which comprise 99% of the atmosphere. Everyone agrees that these gases are transparent to both incoming and outgoing radiation. That means in layman language that these two gases are neither heated by the Sun's incoming radiation, nor by the Earth's outgoing radiation at night. But the air does get warm. How then does the air get warm if not by radiation? This matter is dealt with later in the book. Why does the air get warm from the bottom up? Clearly there can be only one answer. Earth warms atmosphere by Conduction.

In my eighty-four years as a sentient being upon this plane I cannot recall one single day when the weather was exactly the same one day as the next. The weather, the barometric pressure, the winds, the rains, snows or frosts, every single day differs from the next. One year when I lived at Kingston-on-Thames the Thames froze over. Some years Spring has been early but others it seemed that summer would never come. Yet I have friends who get quite hot under the collar, accusing me of not believing in climate change. On the contrary I not only believe in changes of climate, I observe the changing weather patterns every single day. Of course what my Warmist friends mean is that I do not believe in man-made climate change. There they are correct – I certainly do not.

Again, if we are to be logical, we must have recourse to correct definitions. Climate is defined as the average of weather in a particular location. In that case there are as many climates as there are locations. And Meteorologists agree that there are a host of microclimates. Indeed we know that there are some 29 major systems of climate in the world, according to the Koppen Geiger classification. There is no such entity as 'World Climate.'

We have Ministers, Prime Ministers, Presidents, Bishops and all declaring that we must tackle climate change, as if 'climatechange' were a cause and not an effect. But does this declaration bear critical and logical examination? Only a few years ago the eruptions of volcanoes in Iceland caused chaos over Europe, as Air Lines were diverted. Could the said plenipotentiaries prevent a recurrence? Could they in all modesty prevent any single volcanic eruption or tremor in the world, whether say in Italy or Indonesia, in Japan or the United States? Yet these eruptions have a massive and immediate effect on the weather, the average of which is climate.

This winter we were beset by floods in England, which were attributed partly to the Jetstream that travels way up in the Stratosphere. Are we to imagine that any of our Ministers knows some way to alter the Jetstream? Or, does any Minister in India know how to get the Monsoons to arrive one day earlier than Great Nature has decreed?

So we see here quite clearly that we are often not dealing with matters of scientific fact, but spurious opinions based on lack of definitions and a total lack of logic. Anyone who imagines that they can tackle changes of climate is suffering from severe delusions of grandeur. Are we seriously to imagine further that by attempting to limit Carbon Dioxide in the atmosphere that that will reduce warming and hold back 'climate change?' Is that a logical position?

While the Alarmists insist that any ill defined warming is man-made, they all insist on discounting the most obvious source of warmth, which is the Sun. Every child knows that it gets hotter when the Sun shines and cooler when the Sun sets. And there are graphs, which clearly show the variations of TSI, Total Solar Irradiance. The Alarmists discount this yet insist that such variations as we have are caused by man-made additions to a trace gas, which is 0.04% of the atmosphere. And yet it is easy enough to watch documentaries on 'Our Secret Sun' to see the explosions on the Sun's surface, to realise the magnitude of the Sun in relation to our Earth, to realise that there are solar winds and charged particles bombarding us. And all of this mighty force is disregarded in favour of the supposed effects of a trace and beneficial gas.

These essays both by others and myself have been put in chronological order, as they were often triggered by happenings at the time of writing. These essays will inevitably show certain inconsistencies, as my own understanding of the problems advanced, and as the scientists, who became my friends, also advanced and had clearer and clearer intuitions. A good scientist is not static, but is forever enjoying further insights and modifying his or her views

It is to be hoped that this will be at least a blast on the trumpet against the irrational ideas of man-made Global Warming and the illogical idea that man has the power to control the waves, the tectonic plates the numerous submarine volcanoes, the Jetstream, cloud formation, thunder, lightning, floods, tsunamis and other acts of Great Nature.

I dedicate this book to all who have contributed to fighting against this grand delusion. I dedicate this book to the sceptics of this world.

Anthony Bright-Paul
7.07.2014

There IS a problem with global warming... it stopped in 1998

By Bob Carter
Prof Bob Carter is a geologist at James Cook University, Queensland, engaged in paleoclimate research.

Last Updated: 12:01am BST 09/04/2006

For many years now, human-caused climate change has been viewed as a large and urgent problem. In truth, however, the biggest part of the problem is neither environmental nor scientific, but a self-created political fiasco. Consider the simple fact, drawn from the official temperature records of the Climate Research Unit at the University of East Anglia, that for the years 1998-2005 global average temperature did not increase (there was actually a slight decrease, though not at a rate that differs significantly from zero).

Yes, you did read that right. And also, yes, this eight-year period of temperature stasis did coincide with society's continued power station and SUV-inspired pumping of yet more carbon dioxide into the atmosphere.

In response to these facts, a global warming devotee will chuckle and say "how silly to judge climate change over such a short period". Yet in the next breath, the same person will assure you that the 28-year-long period of warming which occurred between 1970 and 1998 constitutes a dangerous (and man-made) warming. Tosh! Our devotee will also pass by the curious additional facts that a period of similar warming occurred between 1918 and 1940, well prior to the greatest phase of world industrialisation, and that cooling occurred between 1940 and 1965, at precisely the time that human emissions were increasing at their greatest.

Does something not strike you as odd here? That industrial carbon dioxide is not the primary cause of earth's recent decadal-scale temperature changes doesn't seem at all odd to many thousands of independent scientists. They have long appreciated - ever since the early 1990s, when the global warming bandwagon first started to roll behind the gravy train of the UN Inter-governmental Panel on Climate Change (IPCC) - that such short-term climate fluctuations are chiefly of natural origin. Yet the public appears to be largely convinced otherwise. How is this possible?

Since the early 1990s, the columns of many leading newspapers and magazines, worldwide, have carried an increasing stream of alarmist letters and articles on hypothetical, human-caused climate change. Each such alarmist article is larded with words such as "if", "might", "could", "probably", "perhaps", "expected", "projected" or "modelled" - and many involve such deep dreaming, or ignorance of scientific facts and principles, that they are akin to nonsense.

The problem here is not that of climate change per se, but rather that of the sophisticated scientific brainwashing that has been inflicted on the public, bureaucrats and politicians alike. Governments generally choose not to receive policy advice on climate from independent scientists. Rather, they seek guidance from their own self-interested science bureaucracies and senior advisers, or from the IPCC itself. No matter how accurate it may be, cautious and politically non-correct science advice is not welcomed in Westminster, and nor is it widely reported.

Marketed under the imprimatur of the IPCC, the bladder-trembling and now infamous hockey-stick diagram that shows accelerating warming during the 20th century - a statistical construct by scientist Michael Mann and co-workers from mostly tree ring records - has been a seminal image of the climate scaremongering campaign. Thanks to the work of a Canadian statistician, Stephen McIntyre, and others, this graph is now known to be deeply flawed.

There are other reasons, too, why the public hears so little in detail from those scientists who approach climate change issues rationally, the so-called climate sceptics. Most are to do with intimidation against speaking out, which operates intensely on several parallel fronts.

First, most government scientists are gagged from making public comment on contentious issues, their employing organisations instead making use of public relations experts to craft carefully tailored, frisbee-science press releases. Second, scientists are under intense pressure to conform with the prevailing paradigm of climate alarmism if they wish to receive funding for their research. Third, members of the Establishment have spoken declamatory words on the issue, and the kingdom's subjects are expected to listen.

On the alarmist campaign trail, the UK's Chief Scientific Adviser, Sir David King, is thus reported as saying that global warming is so bad that Antarctica is likely to be the world's only habitable continent by the end of this century. Warming devotee and former Chairman of Shell, Lord [Ron] Oxburgh, reportedly agrees with another rash statement of King's, that climate change is a bigger threat than terrorism. And goodly Archbishop Rowan Williams, who self-evidently understands little about the science, has warned of "millions, billions" of deaths as a result of global warming and threatened Mr Blair with the wrath of the climate God unless he acts. By betraying the public's trust in their positions of influence, so do the great and good become the small and silly.

Two simple graphs provide needed context, and exemplify the dynamic, fluctuating nature of climate change. The first is a temperature curve for the last six million years, which shows a three-million year period when it was several degrees warmer than today, followed by a three-million year cooling trend which was accompanied by an increase in the magnitude of the pervasive, higher frequency, cold and warm climate cycles. During the last three such warm (interglacial) periods, temperatures at high latitudes were as much as 5 degrees warmer than today's. The second graph shows the average global temperature over the last eight years, which has proved to be a period of stasis.

The essence of the issue is this. Climate changes naturally all the time, partly in predictable cycles, and partly in unpredictable shorter rhythms and rapid episodic shifts, some of the causes of which remain unknown. We are fortunate that our modern societies have developed during the last 10,000 years of benignly warm, interglacial climate. But for more than 90 per cent of the last two million years, the climate has been colder, and generally much colder, than today. The reality of the climate record is that a sudden natural cooling is far more to be feared, and will do infinitely more social and economic damage, than the late 20th century phase of gentle warming.

The British Government urgently needs to recast the sources from which it draws its climate advice. The shrill alarmism of its public advisers, and the often eco-fundamentalist policy initiatives that bubble up from the depths of the Civil Service, have all long since been detached from science reality. Internationally, the IPCC is a deeply flawed organisation, as acknowledged in a recent House of Lords report, and the Kyoto Protocol has proved a costly flop. Clearly, the wrong horses have been backed.

As mooted recently by Tony Blair, perhaps the time has come for Britain to join instead the new Asia-Pacific Partnership on Clean Development and Climate (AP6), whose six member countries are committed to the development of new technologies to improve environmental outcomes. There, at least, some real solutions are likely to emerge for improving energy efficiency and reducing pollution.

Informal discussions have already begun about a new AP6 audit body, designed to vet rigorously the science advice that the Partnership receives, including from the IPCC. Can Britain afford not to be there?

• Prof Bob Carter is a geologist at James Cook University, Queensland, engaged in paleoclimate research

Letter No 1 from Professor Tim Ball
06.04.2007

Dear Tony,

Yes, you have the statistics correct. The first graph is based on calculations produced by atmospheric chemist Fred Singer. The CFCs were added during the so-called ozone 'hole' scare. The fact is they (commercial name is freons) are four times denser than air. I don't know of any mechanism that can put them up in the air and maintain them there. Any refrigerator person knows when released they go straight down. It was argued they are broken down by sunlight and chlorine was the active ingredient for ozone destruction, but then it is no longer a greenhouse gas.

The other point is the argument that though water vapour is by far the most abundant and important greenhouse gas it is not as 'effective' as CO_2 or methane. Some place its effectiveness in trapping heat (long wave radiation) as low as 60%. Most don't agree with this figure but do acknowledge it is not as effective as CO_2 or methane. Despite this, because of its sheer volume, it is by far the most important greenhouse gas. More important it varies considerably naturally from almost zero percent of the atmosphere at the poles to 4% in the tropics. The variation is due to the fact that air's ability to hold water vapour is a function of temperature. In meteorology they have devised several ways of determining water content of the air and ironically the public are only familiar with the worst one - relative humidity.

The percentage of water vapour in the air is determined by the space available between the other molecules. As the air temperature decreases the number of molecules per unit volume decreases, that is there is less space between the air molecules, therefore less space for the water molecules. At a certain point (the point of condensation) the air is cooled so the water vapour changes from a gas to a liquid (water droplets) and clouds are formed. Even more cooling will lead to precipitation. So water vapour is the only gas that varies considerably in the atmosphere in space and time. The amount of water vapour is such that they can determine the amount of atmospheric pressure added by the water vapour content - this is known as the saturation vapour pressure. They also determine the water vapour as a weight per unit volume of a parcel of air or as a ratio of wet to dry air. If this is confusing it is because water vapour is the most important and complex gases in the atmosphere. Equally important is water's ability to exist as a gas, liquid and solid at the same temperature or to change from one form to another as temperature changes - this process is known a Phase change.

Now you know why I want to change the name of the planet from Earth to Water. Beyond its life supporting function it is unique in its properties and crucial to atmospheric processes especially transportation of heat.

Regards
Tim Ball

Letter 2 from Professor Tim Ball

On 8-Apr-07

Re: Facts and the BBC.doc

Tony:

All lakes are geologically temporary features caused by a blockage in the drainage system. Extremely old landscapes have no lakes because the drainage system has evolved fully. Very young landscapes, like those recently out form under glaciations such as most of Canada, have many lakes in what is generally described as a deranged drainage system.

Lake Chad is a remnant lake from the last glacial period and is disappearing, as there is no rainfall to replenish the lake. During the last ice age the subtropical regions experienced increased precipitation during glacial periods and decreased precipitation during interglacials. These cycles are called pluvials (the wet periods) and interpluvials (the dry periods). We are in an interglacial so the subtropics are experiencing interpluvials.

This practice of the BBC and most other media is one of picking on an issue then drawing completely wrong conclusions. Sadly, they can find specialists in that one area who give them quote so buttress their position, but when you put it in the full context you discover there are other perfectly logical explanations.

Tim Ball

Letter No 3 from Professor Tim Ball

08.07.2007

Hi Tony:

It is reasonable to argue that at 385 ppm CO_2 is very low for plant viability. Research shows they operate better between 1000 and 1200 ppm, which suggests they evolved to that level. Commercial greenhouses are injecting up to 1000 ppm and the plants grow four times larger and with greater yields while using less water. This suggests the plants evolved to that level and are now essentially CO_2 starved (as Shaviv's comments attest). Reducing the atmospheric level, as governments plan, will be harmful to the plants.

The permafrost melting was not included for a variety of reasons including not enough time. More important the story was debunked by Russian permafrost experts. The argument is the melting would release methane, but atmospheric levels of methane have declined for at least 14 years. Of course, no apologies to the farmers who were accused through their cows of being responsible for the methane and global warming. You can go back through the literature and see that virtually every 'scare' such as thin eggshells, deformed frogs, coral bleaching are now completely proven incorrect in their cause and condition.

I am sorry I can't help you with the DVD. I am still waiting for a copy from the producer myself, but I know he is extremely busy. The other issue the program didn't receive government funding so has to recover costs from sales.

Regards
Tim Ball

November 1, 2007

My Nobel Moment

By JOHN R. CHRISTY

I've had a lot of fun recently with my tiny (and unofficial) slice of the 2007 Nobel Peace Prize awarded to the Intergovernmental Panel on Climate Change (IPCC). But, though I was one of thousands of IPCC participants, I don't think I will add "0.0001 Nobel Laureate" to my resume.

The other half of the prize was awarded to former Vice President Al Gore, whose carbon footprint would stomp my neighborhood flat. But that's another story.

Both halves of the award honor promoting the message that Earth's temperature is rising due to human-based emissions of greenhouse gases. The Nobel committee praises Mr. Gore and the IPCC for alerting us to a potential catastrophe and for spurring us to a carbonless economy.

I'm sure the majority (but not all) of my IPCC colleagues cringe when I say this, but I see neither the developing catastrophe nor the smoking gun proving that human activity is to blame for most of the warming we see. Rather, I see a reliance on climate models (useful but never "proof") and the coincidence that changes in carbon dioxide and global temperatures have loose similarity over time.

There are some of us who remain so humbled by the task of measuring and understanding the extraordinarily complex climate system that we are skeptical of our ability to know what it is doing and why. As we build climate data sets from scratch and look into the guts of the climate system, however, we don't find the alarmist theory matching observations. (The National Oceanic and

Atmospheric Administration satellite data we analyze at the University of Alabama in Huntsville does show modest warming -- around 2.5 degrees Fahrenheit per century, if current warming trends of 0.25 degrees per decade continue.)

It is my turn to cringe when I hear overstated-confidence from those who describe the projected evolution of global weather patterns over the next 100 years, especially when I consider how difficult it is to accurately predict that system's behavior over the next five days. Mother Nature simply operates at a level of complexity that is, at this point, beyond the mastery of mere mortals (such as scientists) and the tools available to us. As my high-school physics teacher admonished us in those we -shall-conquer-the-world-with-a-slide-rule days, "Begin all of your scientific pronouncements with 'At our present level of ignorance, we think we know . . .'" I haven't seen that type of climate humility lately. Rather I see jump-to-conclusions advocates and, unfortunately, some scientists who see in every weather anomaly the specter of a global -warming apocalypse. Explaining each successive phenomenon as a result of human action gives them comfort and an easy answer.

Others of us scratch our heads and try to understand the real causes behind what we see. We discount the possibility that *everything* is caused by human actions, because everything we've seen the climate do has happened before. Sea levels rise and fall continually. The Arctic ice cap has shrunk before. One millennium there are hippos swimming in the Thames, and a geological blink later there is an ice bridge linking Asia and North America.

One of the challenges in studying global climate is keeping a global perspective, especially when much of the research focuses on data gathered from spots around the globe. Often observations from one region get more attention than equally valid data from another.
The recent CNN report "Planet in Peril," for instance, spent considerable time discussing shrinking Arctic sea ice cover. CNN did *not* note that winter sea ice around Antarctica last month set a record maximum (yes, maximum) for coverage since aerial measurements started.

Then there is the challenge of translating global trends to local climate. For instance, hasn't global warming led to the five-year drought and fires in the U.S. Southwest? Not necessarily. There has been a drought, but it would be a stretch to link this drought to carbon dioxide. If you look at the 1,000-year climate record for the western U.S. you will see not five-year but 50-year-long droughts. The 12th and 13th centuries were particularly dry. The inconvenient truth is that the last century has been fairly benign in the American West. A return to the region's long-term "normal" climate would present huge challenges for urban planners.

Without a doubt, atmospheric carbon dioxide is increasing due primarily to carbon- based energy production (with its undisputed benefits to humanity) and many people ardently believe we must "do something" about its alleged consequence, global warming. This might seem like a legitimate concern given the potential disasters that are announced almost daily, so I've looked at a couple of ways in which humans might reduce CO_2 emissions and their impact on temperatures.

California and some Northeastern states have decided to force their residents to buy cars that average 43 miles-per-gallon within the next decade. Even if you applied this law to the entire world, the net effect would reduce projected warming by about 0.05 degrees Fahrenheit by 2100, an amount so minuscule as to be undetectable. Global temperatures vary more than that from day to day.

Suppose you are very serious about making a dent in carbon emissions and could replace about 10% of the world's energy sources with non-CO_2-emitting nuclear power by 2020 -- roughly equivalent to halving U.S. emissions. Based on IPCC-like projections, the required 1,000 new nuclear power plants would slow the warming by about 0.2176 degrees Fahrenheit per century. It's a dent.

But what is the economic and human price, and what is it worth given the scientific uncertainty? My experience as a missionary teacher in Africa opened my eyes to this simple fact: Without access to energy, life is brutal and short. The uncertain impacts of global warming far in the future must be weighed against disasters at our doorsteps today. Bjorn Lomborg's Copenhagen Consensus 2004, a cost-benefit analysis of health issues by leading economists (including three Nobelists), calculated that spending on health issues such as micronutrients for children, HIV/AIDS and water purification has benefits 50 to 200 times those of attempting to marginally limit "global warming."

Given the scientific uncertainty and our relative impotence regarding climate change, the moral imperative here seems clear to me.

Mr. Christy is director of the Earth System Science Center at the University of Alabama in
Huntsville and a participant in the U.N.'s Intergovernmental Panel on Climate Change,
co-recipient of this year's Nobel Peace Prize.
URL for this article:
http://online.wsj.com/article/SB119387567378878423.html

Letter No 4 from Professor Tim Ball

16.12.2007

Hi Anthony,

The article looks fine to me. Here are some comments about CO2 and some web sites with more information. One important thing to understand is the education system has, since Darwin, assumed change is gradual over long periods. It is called uniformitarianism and if it sounds religious it is because it is a belief that underpins how we view the world. As your article points out, change is significant in time and magnitude.

Here are some comments about CO2 and some good sites for more information.

It is difficult to find contradictory science. There are several reasons for this. First, the hypothesis that human addition of CO2 to the atmosphere would lead to global warming was accepted as fact almost as soon as it was produced. As Lindzen said about 20 years ago, The consensus was reached before the research had even begun. As soon as scientists moved to challenge the hypothesis following the normal scientific method set out by Kuhn, they were attacked as Skeptics. In fact all scientists should be Skeptics. More recently we have been called "deniers" with all the holocaust connotations of that word. This effectively thwarted the scientific method.

Second, the problem was exacerbated; as billions in research funds from governments were directed to proving not disproving the hypothesis as Karl Popper said is essential to scientific inquiry.

Third, scientific journals did not concern themselves with climate until the funding went that way and it became a political issue. Editors sent articles questioning the prevailing wisdom to the high priests who reject it as heresy. I use religious terminology because what is happening is not science. I call it peer review censorship.

Despite these efforts a few scientists have struggled on and produced a great deal of evidence that the hypothesis is wrong. I have all too briefly summarized this for you. Fourth, almost all the funding goes to proving the hypothesis.

A common problem with this entire issue of climate change is confusion between pollution and global warming or climate change.. Despite attempts to muddy the water, for example by the Canadian government listing CO2 as a toxic substance, **global warming and climate change are not about pollution**. Questioning the climate science does not mean there is not a recognition of or concern about clean air and water and a healthier environment. CO2 is not a pollutant, it is a naturally occurring gas. Indeed, it is reasonable to argue that reducing CO2 is a negative move.
Some salient points about CO2

It is at 385 ppm (bottom right side of diagram) at the lowest level in 600 million years.

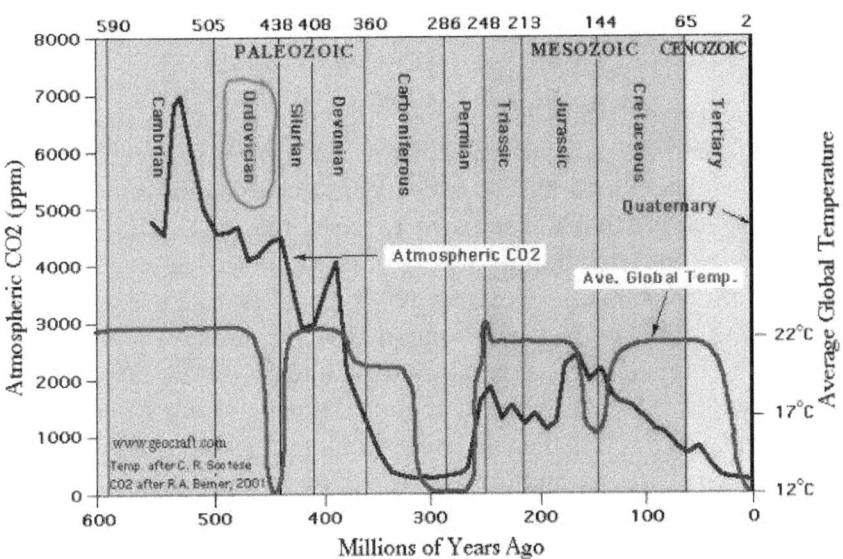

Notice there is no correlation between CO2 and temperature at any point. Research by Sherwood and Craig Idsohttp://www.co2science.org/ shows most plants function best between 1000 and 1200 ppm. Commercial greenhouses are pumping these amounts in and achieving four times better growth and yield with significantly less water use. This suggests the plants have evolved to that level and our now CO2 'starved.'

2. Concern that temperatures will rise with increasing CO2 are cancelled by a couple of important facts. First, the ice core covering 420,000 years shows that temperature in creases before CO2. This is in complete contradiction to the assumption that an increase in CO2 due to humans will cause an increase in temperature.

3. Even if CO2 doubles or triples there is an upper limit to the amount of temperature increase that can occur because of its role as a greenhouse gas. This upper limit is set at 1.5 C, but some like Richard Lindzen believe it is less. The reason for this upper limit is because the atmosphere is close to saturation for CO2 now, so any further increase will not increase its ability to 'trap' heat. A good analogy is painting a window with black paint to block the sunlight. The first coat achieves 98% reduction. Any subsequent coats achieve dramatically less reduction.

4. They get round this last problem in the computer models by assuming with no scientific justification a 'positive' feedback. A positive feedback in climate science means one that enhances a trend, while a negative feedback stops or reverses the trend. They assume the increase in temperature will result in increased evaporation, which will cause more heat to be trapped and temperatures to continue to rise. In fact, the amount of increased evaporation is negligible compared to the total amount already in the atmosphere. Water Vapour is by far the most abundant greenhouse gas being 95% by volume. CO2 is less than 4% by volume and the human portion of that is about 0.4%. In addition, the increased water vapour is more likely to increase cloud cover which will act as a negative feedback blocking the sun and causing cooling.

A few sources you can follow include our own web page, http://www.nrsp.com/

I also helped set up and remain an advisor behind the scenes for Friends of Science at http://www.friendsofscience.org/

Two other sites I recommend include a more scientific one at with much material on the global average annual temperature http://www.co2science.org/

If you want more in depth discussion about the manipulation of data visit the site of Steve McInytyre who exposed the "hockey stick" fraud. You will find extensive discussion here on the latest "adjustments to the temperature record, which, in my opinion border on malfeasance. http://www.climateaudit.org/ and a more general one at http://www.junkscience.com/

You can use the links provided on these sites to lead you to other material
Here are ten excellent newspaper articles with good information.
http://www.canada.com/nationalpost/story.html?id=22003a0d-37cc-4399-8bcc-39cd20bed2f6&k=0

Tolstoy had a comment on the conundrum of several scientists refusing to acknowledge what the evidence show. Sadly, a vast majority simply don't know climate science.

"I know that most men, including those at ease with problems of the greatest complexity, can seldom accept even the simplest and most obvious truth if it be such as would oblige them to admit the falsity of conclusions which they delighted in explaining to colleagues, which they have proudly taught to others, and which they have woven, thread by thread, into the fabric of their lives."

I am not quoting it correctly, but Ghandi's four stages of change are first they ignore you, then they laugh at you, then they attack and then you succeed. It appears we are at stage three.

Regards

Tim Ball

Causes of Climate Change

Anthony Bright-Paul
December 17th 2007.

Until such time as we know the causes of climate change it is useless and indeed arrogant for human beings, however well intentioned, to imagine that they can tackle climate change. We know that climate has been changing for some 450,000 years, as one Glacial Period (Ice Age) is succeeded by Interglacials (Warm Periods).

Such a Warm Period occurred relatively recently, between 900 and 1300 A.D. and is known as the Mediaeval Warm Period. This was the time when the Norsemen colonised Iceland and the southern quarter of Greenland, sailing in open boats round Scotland to the Isle of Man.

The Little Ice Age followed for some 400 years, during which time the Thames regularly froze over. Technically we are just emerging from this period.

The cause of these temperature variations is not completely known, but we do know what is called the Milankovitch effect, when the earth's orbit changes from almost circular to extreme ellipse round the
sun (the tilt changes from 21.8° to 24.4° and the date of the equinox varies), so when we get nearer the sun, we get warmer. This is not rocket science – everyone knows that they get warm when the sun shines.

As the earth warms up, the oceans (the major source of CO2) get warmer, sending streams of carbon dioxide into the atmosphere. Mr Al Gore perhaps conveniently failed to notice that the temperature rises <u>before</u> the CO2.

What do we know about Carbon Dioxide? Firstly it is a natural gas, **odourless, colourless** and not a pollutant. We humans breathe in oxygen and breathe out carbon dioxide, all 6.2 billion of us, night and day. So do all living organisms. Decaying vegetation sends streams of CO2 into the air. This CO2 is continually re-cycled through plants and through being dissolved in the colder oceans. Indeed, there is a constant movement of CO2 between the oceans and the atmosphere that varies with the water temperature.

The Greenhouse Gases are composed of 95% Water Vapour. Carbon Dioxide is less than 4% from all sources, the anthropogenic or man-made portion being less than 0.117% of that 4%. The idea that CO2 drives climate simply does not hold water.

Much confusion arises because of those pictures we all see of smoking chimneystacks, hurling Sulphur Dioxide and particulate matter into the air. That indeed is pollution and that is something that can be dealt with by humans here on earth. This smoke can be 'scrubbed'. London is a wonderful example of how the terrible pea-souper fogs of my childhood have now been changed to a city of excellent air.

Likewise the River Thames was virtually dead from chemical pollutants, but is now once again teeming with fish. We have also learned how to protect our coastlines from oil spillage and international laws attempt to keep the seas clean for our wild life and for the underwater coral reefs.

It is right and proper that we should all do all we can to guard our environment and in this respect there cannot be a sentient being that is not 'green.' We are all at one in this regard.

But when it comes to *tackling climate change* there we humans are at a loss. We cannot change the climate, for we cannot even change the weather. At this very moment Australia is suffering a drought. No man has yet become a rainmaker. We can build buildings that protect to a large degree from storms and hurricanes, but though we can predict them, in no way can we stop them in their course. In the face of Great Nature man can only pray to the Powers that be. Monsoons will come and go. The tectonic plates will shift, tsunamis will arrive, volcanoes will erupt and there is absolutely nothing that man can do except to take as adequate precautions as possible against their eventuality.

Such is the pre-occupation with carbon and carbon footprints, stoked up by the media and by governmental funded bodies, that we have taken our eye off the ball on the major problem of world poverty. Of the six billion people on this earth's surface one third of them, that is some two billion people, are living a life of poverty amidst a world of incredible riches. What is the difference between the developed world and the Third World?

In one word it is Electricity.

Without electricity, we cannot have central heating or air conditioning in our homes; without electricity we can have neither running tap water or sewage systems; without electricity we cannot have refrigerators to preserve food or medicines; without electricity we can have no light at night, nor could we run computers, telephones, mobile phones; and above all without electricity we can have neither scientific research nor adequate hospitals.

The distinguishing feature of the Third World is not Geographical, though that has some bearing, but lack of electricity. For those of us in the Western World who hold conferences with laptops in air conditioned hotels, the condition of those who live still in smoke filled hovels with no light, no running water, no sewage systems, no means of communication, is inconceivable.

It is impossible for us by means of donations to charities alone to meet all the demands of the poor. Clearly we should do all we can to help the Third World to get hold of the cheapest forms of power possible. Unfortunately wind-power though desirable in theory, is simply too expensive. Where the world is poor they need the cheapest possible forms of generating power.

If the Globe has warmed half a degree centigrade in the last hundred years and we are entering a Warm Period, let us rejoice without fear. Let us pay no heed to the scaremongers and the politicians, and accept that we are entering a glorious period for the whole world and the biosphere. Don't let us hinder it with false, needless and unscientific predictions of catastrophe.

Anthony Bright-Paul

December 17th 2007.

Letter No 5 from Professor Tim Ball

04.02.2008

Hi Anthony:

Thanks for your kind words and I am not surprised Professors Carter and Christy reply - they are honourable people genuinely seeking the truth and I am sure like me if they found evidence to contradict their position they would be the first to acknowledge its existence.

Your concern about the propaganda war is one many hold and understandable in the great noise and efficiency of technique employed. If you want to sway people's emotions propaganda can do it better than cold hard scientific facts every time. As I point out it is difficult to counteract Gore's movie in the classroom because of these inherent differences.

I am not as pessimistic as you because I have struggled with the role and function of extremists in society. Every new paradigm brings change and adherents. The majority see the values but also fear the changes and have a basic conservatism. This was true of the new paradigm of environmentalism. Certainly we knew it made basic good and common sense to act responsibly within our home.

However, as with all paradigms a few saw other opportunities, chances to impose their political and moral views and values on others. As H L Mencken said many years ago, "The urge to save humanity is almost always a false front for the urge to rule." These people are the fourth standard deviation of the spectrum offset at the other end by those who don't want to change at all. The vast majority are in the middle saying we must change but how far do we go. The extremists start acting and doing more and more extreme things and in so doing provide the limits for the rest. That is happening right now. Increasingly extreme statements and claims gradually make even those who don't understand the science look up and say, no, now we are losing more than we are gaining. Statements such as "The science is settled" raised eyebrows among the media and many of the public who know the science is never settled. The switch of focus from global warming to climate change to offset the fact very slight cooling is occurring while CO_2 levels continue to rise in contradiction to the fundamental assumption of the hypothesis that humans are the cause. Now they claim warming, cooling, wetter, drier, more storms, fewer storms and any other event is due to human activity strikes the chord of illogic.

Another year of cooling (and many expect several years as sunspot activity declines) will further underscore the political nature of the issue.

Charles Mackay said, "Men, it has been well said, think in herds; it will be seen that they go mad in herds, while they only recover their senses slowly, and one by one." I don't agree. Why would the herd mentality disappear? What will happen is the herd will simply change direction. My concern for some time is the herd will say about science, "We don't believe anything you tell us," and real issues won't be addressed.

Thanks for your willingness to get the all facts and then educate others to both sides of the issue.

Regards
Tim Ball

The sea is a fizzy drink!

Anthony Bright-Paul

March 23rd 2008.

We all love to watch the sea, particularly when it is rough and breakers crash upon the rocks sending spray into the air. This is Great Natures way of dissolving carbon dioxide in its waters. Like Coca-cola, like Champagne, the sea is a vast fizzy drink.

If you get a bottle of Ginger Ale and heat it lightly on a stove, very quickly the bubbles will disappear and the fizzy drink will go flat. In exactly the same way, when the sun warms the oceans, the bubbles, the fizz, evaporates into the air and rises up in streams. Can you see it? No alas! You cannot. Carbon Dioxide is colourless and odourless, and is certainly not a pollutant.

Are you puzzled? I know why. It is probably because you have seen that Science Fiction film of Vice President Al Gore. Or you may even have seen pictures repeated on the BBC of fearful power stations with huge chimneys spouting black foul-smelling, sulphurous smoke into the sky. You have been lead to believe that that foul black stuff, and the foul exhausts from car and aeroplanes is Carbon Dioxide. My dear, it is a scam. It is a lie.

How can I be so certain? Easily. Everybody knows that we breathe. You know that you breathe. With the most elementary learning you know that you breathe in Oxygen and you exhale Carbon Dioxide. All the 6.2 billions of us on the earth, night and day, are exhaling Carbon Dioxide. As well as all the animals of the animal kingdom – you will excuse me if I have been unable to count them!

Do you see humans snorting black smoke? If you do, look out! That's not a human, you have met a dragon! If you want to prove it to yourself blow up a kid's balloon – in this way you can capture some Carbon Dioxide of your own. Pretty harmless stuff, wouldn't you say?

But then you will ask, what is all this stuff about Carbon Footprints, Global Warming, Climate Change and Saving the Planet? What indeed! You are a witness to the greatest scam, the most incredible hoax that has ever been perpetrated on Mother Earth.

You have been lead to believe by an incessant and relentless propaganda campaign that the whole Globe is Warming at such a rate, that glaciers are melting, that the Arctic Ice-Cap will probably melt within 5 years! and that will cause such havoc that Manhattan will flood, let alone half of Florida and Lord alone knows what will happen to India. If you believe all of this you are not alone. Even Prime Ministers, even Chancellors, Secretary Generals, Archbishops, even some quite eminent scientists of different disciplines, have been caught up in the hysteria created by this monumental hoax. The voices of Doom are telling you that the tipping point is approaching, that Climate Change is going to change the planet, and moreover you have got to act now. Above all you must not fly! Why? Because you will be using up your ration of Carbon!

Rubbish! Complete twaddle! In this way you are being asked to give up your greatest gift, which is Common Sense.

You do not have to be a Rocket Scientist to know that the Sun warms the earth. Even little Susie knows she get warm when the sun shines. And when little Susie grows up she soon learns that heat rises. That is not difficult is it? You don't have to be a scientist to understand something so simple.

Actually as far as the Globe is concerned you might be surprised to know that for the last eight years the Globe has been cooling, not warming at all! Personally I am disappointed, as I was hoping for an opening of the North-West Passage.

But what about the Greenhouse Effect? Isn't that making us all a whole lot hotter and sending the planet haywire? Not at all, once again we have been fed with a mish-mash of twaddle. Without the Greenhouse Gases the whole earth would be miles too hot during the day and miles too cold during the night – like living in the Moon. Life on earth would be impossible. I know what you are thinking. Some bright spark has put up the idea that Carbon Dioxide is accumulating up there against some imaginary glass panel up in the Troposphere, and that is causing the trouble. Wrong again! The Greenhouse Gases are mainly Water Vapour, 95%. And Carbon Dioxide? Of yes, there is Carbon Dioxide. It is a teeny-weeny, absolutely tiny trace element, and it is guilty of nothing at all.

On the contrary at the present time we are CO2 impoverished. We need CO2 as it is a plant nutrient. It is true that we need some 2 billion acres more trees to provide us with oxygen, but those trees in turn need Carbon Dioxide!!

Have a drink of Champagne, my dear, and put on my favourite song "Don't you worry about a thing!"

Scientific Consensus

Anthony Bright-Paul
03.05.2008

I have been excommunicated by one of my friends as he upbraided me for not paying due respects to the Scientific Consensus. Oh, woe is me! No more will he hear my ravings!

But pause for a moment – what is this so-called 'scientific consensus?' He was of course referring to the current hullabaloo about Global Warming and Climate Change – something that I with enormous arrogance have dared to question. Why, everybody knows that we humans are the great polluters, and Vice President Al Gore shows a map glowing red, showing that the United States, of which he is a citizen, are the worst polluters. I have just watched his film once again – it's magic! (Black Magic!)

It is by no means difficult to get an 'emotional consensus'. If I were in a room filled with the most hardened Skeptics, and asked everyone to raise their hands that was in favour of 'pollution', not one hand would be raised. If I were to ask in any congregation anywhere in the world, 'Who is in favour of saving this Planet? Who is against pollution? Who is in favour of saving the Siberian White tigers? The minke whales? The cuddly Polar Bears? The Amazonian forests?' Why! Every single person would raise their hands. Every single person that I know is in favour of looking after the environment. And not only individuals, but also local governments of all descriptions are in favour of the Environment – that is nothing new.

But what is new is a form of censorship that has grown up – everybody knows that it is generally accepted – anthropogenic carbon dioxide is causing Global Warming – Global Warming is causing Climate Change – we must reduce our Carbon Footprints – we must save the Polar Bears – Global Warming will cause the Himalayan Glaciers to melt with untold suffering to the Indian sub-continent as huge flooding occurs. One scenario of disaster follows another. In fact the whole business is kept going by hysteria about forthcoming disasters, the melting ice-caps, the tipping points beyond which this Planet, this beloved Planet of ours, would not be able to recover.

In order to keep this scenario going, in order to keep it fresh in everybody's minds, Vice President Al Gore stomps the world with his 'Inconvenient Truth', while his acolytes do everything possible to stifle or rubbish any dissent. So effective has this ploy been, it is no great wonder that many people believe that there is a Scientific Consensus.

Nowadays, before we accept anything we want proof. I am going to prove to you, my reader, that is, if you will continue to read, I am going to prove to you that no such Scientific Consensus exists. There is an emotional and often hysterical consensus, but **there is no Scientific Consensus**, in spite of what you may have been lead to believe.

We are told by the IPCC that we must address the issue of Climate Change. So strong is this advice that Governments bow to their edicts, Chancellors pronounce on them, Prince Charles, the Archbishop of Canterbury and now even Tony Blair, get in on the act. The case for Scientific Consensus seems overwhelming. Except for one thing.

At the Conference in Manhattan (you were not meant to know about this) 500 scientists were signatories to the Manhattan Declaration.

Manhattan Declaration on Climate Change

"Global warming" is not a global crisis

We, the scientists and researchers in climate and related fields, economists, policymakers, and business leaders, assembled at Times Square, New York City, participating in the 2008 International Conference on Climate Change,

Resolving that scientific questions should be evaluated solely by the scientific method;

Affirming that global climate has always changed and always will, independent of the actions of humans, and that carbon dioxide (CO_2) is not a pollutant but rather a necessity for all life;

Recognising that the causes and extent of recently observed climatic change are the subject of intense debates in the climate science community and that oft-repeated assertions of a supposed 'consensus' among climate experts are false;

Affirming that attempts by governments to legislate costly regulations on industry and individual citizens to encourage CO_2 emission reduction will slow development while having no appreciable impact on the future trajectory of global climate change. Such policies will markedly diminish future prosperity and so reduce the ability of societies to adapt to inevitable climate change, thereby increasing, not decreasing, human suffering;

Noting that warmer weather is generally less harmful to life on Earth than colder:

Hereby declare:

That current plans to restrict anthropogenic CO_2 emissions are a dangerous misallocation of intellectual capital and resources that should be dedicated to solving humanity's real and serious problems.

***That* there is no convincing evidence that CO_2 emissions from modern industrial activity has in the past, is now, or will in the future cause catastrophic climate change.**

That attempts by governments to inflict taxes and costly regulations on industry and individual citizens with the aim of reducing emissions of CO_2 will pointlessly curtail the prosperity of the West and progress of developing nations without affecting climate.

That adaptation as needed is massively more cost-effective than any attempted mitigation and that a focus on such mitigation will divert the attention and resources of governments away from addressing the real problems of their peoples.

That human-caused climate change is not a global crisis.

Now, therefore, we recommend --

That world leaders reject the views expressed by the United Nations Intergovernmental Panel on Climate Change as well as popular, but misguided works such as "An Inconvenient Truth."

That all taxes, regulations, and other interventions intended to reduce emissions of CO2 be abandoned forthwith.

Agreed at New York, 4 March 2008

Hey! Hey! Hey! Who are these guys? What is all this about CO2 not being a pollutant, but a necessity of life? Are they crazy? Why, this is absolutely contrary to what Archbishop Al Gore has been preaching!

I said I would give you proof. Now here is the proof, and I am sorry that it might take you a bit of time to read this proof and to open your eyes. I want you not only to look at this list, taken from Wikipedia, but also look at their qualifications, at the calibre of these men.

[edit] Believe global warming is not occurring or has ceased

Timothy F. Ball, former Professor of Geography, University of Winnipeg: "[The world's climate] warmed from 1680 up to 1940, but since 1940 it's been cooling down. The evidence for warming is because of distorted records. The satellite data, for example, shows cooling." (November 2004)[5] "There's been warming, no question. I've never debated that; never disputed that. The dispute is, what is the cause. And of course the argument that human CO_2 being added to the atmosphere is the cause just simply doesn't hold up..." (May 18, 2006; at 15:30 into recording of interview)[6] "The temperature hasn't gone up. ... But the mood of the world has changed: It has heated up to this belief in global warming." (August 2006)[7] "Temperatures declined from 1940 to 1980 and in the early 1970's global cooling became the consensus. ... By the 1990's temperatures appeared to have reversed and Global Warming became the consensus. It appears I'll witness another cycle before retiring, as the major mechanisms and the global temperature trends now indicate a cooling." (Feb. 5, 2007)[8]

Robert M. Carter, geologist, researcher at the Marine Geophysical Laboratory at James Cook University in Australia: "the accepted global average temperature statistics used by the Intergovernmental Panel on Climate Change show that no ground-based warming has occurred since 1998 ... there is every doubt whether any global warming at all is occurring at the moment, let alone human-caused warming."[9]

Vincent R. Gray, coal chemist, climate consultant, founder of the New Zealand Climate Science Coalition: "The two main 'scientific' claims of the IPCC are the claim that 'the globe is warming' and 'Increases in carbon dioxide emissions are responsible'. Evidence for both of these claims is fatally flawed."[10]

[edit] Believe accuracy of IPCC climate projections is inadequate

Individuals in this section conclude that it is not possible to project global climate accurately enough to justify the ranges projected for temperature and sea-level rise over the next century. They do not conclude specifically that the current IPCC projections are either too high or too low, but that the projections are likely to be inaccurate due to inadequacies of current global climate modeling.

David Bellamy, environmental campaigner, broadcaster and former botanist: a doubling of atmospheric CO_2 "will amount to less than 1°C of global warming [and] such a scenario is unlikely to arise given our limited reserves of fossil fuels—certainly not before the end of this century."[11]

Hendrik Tennekes, retired Director of Research, Royal Netherlands Meteorological Institute: "The blind adherence to the harebrained idea that climate models can generate 'realistic' simulations of climate is the principal reason why I remain a climate skeptic. From my background in turbulence I look forward with grim anticipation to the day that climate models will run with a horizontal resolution of less than a kilometer. The horrible predictability problems of turbulent flows then will descend on climate science with a vengeance."[12]

Antonino Zichichi, emeritus professor of physics at the University of Bologna and president of the World Federation of Scientists : "models used by the Intergovernmental Panel on Climate Change (IPCC) are incoherent and invalid from a scientific point of view".[13]

[edit] Believe global warming is primarily caused by natural processes

Individuals in this section conclude that the observed warming is more likely attributable to natural causes than to human activities.

Khabibullo Abdusamatov, mathematician and astronomer at Pulkovskaya Observatory of the Russian Academy of Sciences: "Global warming results not from the emission of greenhouse gases into the atmosphere, but from an unusually high level of solar radiation and a lengthy - almost throughout the last century - growth in its intensity...Ascribing 'greenhouse' effect properties to the Earth's atmosphere is not scientifically substantiated...Heated greenhouse gases, which become lighter as a result of expansion, ascend to the atmosphere only to give the absorbed heat away."[14][15][16]

Sallie Baliunas, astronomer, Harvard-Smithsonian Center for Astrophysics: "[T]he recent warming trend in the surface temperature record cannot be caused by the increase of human-made greenhouse gases in the air."[17]

Reid Bryson, emeritus professor of Atmospheric and Oceanic Sciences, University of Wisconsin-Madison: "It's absurd. Of course it's going up. It has gone up since the early 1800s, before the Industrial Revolution, because we're coming out of the Little Ice Age, not because we're putting more carbon dioxide into the air."[18]

George V. Chilingar, Professor of Civil and Petroleum Engineering at the University of Southern California: "The authors identify and describe the following global forces of nature driving the Earth's climate: (1) solar radiation ..., (2) outgassing as a major supplier of gases to the World Ocean and the atmosphere, and, possibly, (3) microbial activities ... . The writers provide quantitative estimates of the scope and extent of their corresponding effects on the Earth's climate [and] show that the human-induced climatic changes are negligible."[19]

Ian Clark, hydrogeologist, professor, Department of Earth Sciences, University of Ottawa: "That portion of the scientific community that attributes climate warming to CO_2 relies on the hypothesis that increasing CO_2, which is in fact a minor greenhouse gas, triggers a much larger water vapour response to warm the atmosphere. This mechanism has never been tested scientifically beyond the mathematical models that predict extensive warming, and are confounded by the complexity of cloud formation - which has a cooling effect. ... We know that [the sun] was responsible for climate change in the past, and so is clearly going to play the lead role in present and future climate change. And interestingly... solar activity has recently begun a downward cycle."[20]

David Douglass, solid-state physicist, professor, Department of Physics and Astronomy, University of Rochester: "The observed pattern of warming, comparing surface and atmospheric temperature trends, does not show the characteristic fingerprint associated with greenhouse warming. The inescapable conclusion is that the human contribution is not significant and that observed increases in carbon dioxide and other greenhouse gases make only a negligible contribution to climate warming."[21]

Don Easterbrook, emeritus professor of geology, Western Washington University: "global warming since 1900 could well have happened without any effect of CO2. If the cycles continue as in the past, the current warm cycle should end soon and global temperatures should cool slightly until about 2035"[22]

William M. Gray, Professor Emeritus and head of The Tropical Meteorology Project, Department of Atmospheric Science, Colorado State University: "This small warming is likely a result of the natural alterations in global ocean currents which are driven by ocean salinity variations. Ocean circulation variations are as yet little understood. Human kind has little or nothing to do with the recent temperature changes. We are not that influential."[23] "I am of the opinion that [global warming] is one of the greatest hoaxes ever perpetrated on the American people."[24] "So many people have a vested interest in this global-warming thing—all these big labs and research and stuff. The idea is to frighten the public, to get money to study it more."[25]

William Kininmonth, meteorologist, former Australian delegate to World Meteorological Organization Commission for Climatology: "There has been a real climate change over the late nineteenth and twentieth centuries that can be attributed to natural phenomena. Natural variability of the climate system has been underestimated by IPCC and has, to now, dominated human influences."[26]

George Kukla, retired Professor of Climatology at Columbia University and Lamont-Doherty Earth Observatory, said in an interview: "What I think is this: Man is responsible for a PART of global warming. MOST of it is still natural."[27]

David Legates, associate professor of geography and director of the Center for Climatic Research, University of Delaware: "About half of the warming during the 20th century occurred prior to the 1940s, and natural variability accounts for all or nearly all of the warming."[28]

Marcel Leroux, former Professor of Climatology, Université Jean Moulin: "The possible causes, then, of climate change are: well-established orbital parameters on the palaeoclimatic scale, ... solar activity, ...; volcanism ...; and far at the rear, the greenhouse effcct, and in particular that caused by water vapor, the extent of its influence being unknown. These factors are working together all the time, and it seems difficult to unravel the relative importance of their respective influences upon climatic evolution. Equally, it is tendentious to highlight the anthropic factor, which is, clearly, the least credible among all those previously mentioned."[29]

Tad Murty, oceanographer; adjunct professor, Departments of Civil Engineering and Earth Sciences, University of Ottawa: global warming "is the biggest scientific hoax being perpetrated on humanity. There is no global warming due to human anthropogenic activities. The atmosphere hasn't changed much in 280 million years, and there have always been cycles of warming and cooling. The Cretaceous period was the warmest on earth. You could have grown tomatoes at the North Pole"[30]

Tim Patterson[31], paleoclimatologist and Professor of Geology at Carleton University in Canada: "There is no meaningful correlation between CO_2 levels and Earth's temperature over this [geologic] time frame. In fact, when CO_2 levels were over ten times higher than they are now, about 450 million years ago, the planet was in the depths of the absolute coldest period in the last half billion years. On the basis of this evidence, how could anyone still believe that the recent relatively small increase in CO_2 levels would be the major cause of the past century's modest warming?"[32][33]

Ian Plimer, Professor of Mining Geology, The University of Adelaide: "We only have to have one volcano burping and we have changed the whole planetary climate... It looks as if carbon dioxide actually follows climate change rather than drives it".[34]

Tom Segalstad, head of the Geological Museum at the University of Oslo: "It is a search for a mythical CO2 sink to explain an immeasurable CO2 lifetime to fit a hypothetical CO2 computer model that purports to show that an impossible amount of fossil fuel burning is heating the atmosphere. It is all a fiction".[35][36]

Nir Shaviv, astrophysicist at the Hebrew University of Jerusalem: "[T]he truth is probably somewhere in between [the common view and that of skeptics], with *natural causes* probably being more important over the past century, whereas *anthropogenic causes* will probably be more dominant over the next century. ... [A]bout 2/3's (give or take a third or so) of the warming [over the past century] should be attributed to increased solar activity and the remaining to anthropogenic causes." His opinion is based on some proxies of solar activity over the past few centuries.[37]

Fred Singer, Professor emeritus of Environmental Sciences at the University of Virginia: "The greenhouse effect is real. However, the effect is minute, insignificant, and very difficult to detect."[38][39] "It's not automatically true that warming is bad, I happen to believe that warming is good, and so do many economists."[40]

Willie Soon, astrophysicist, Harvard-Smithsonian Center for Astrophysics: "[T]here's increasingly strong evidence that previous research conclusions, including those of the United Nations and the United States government concerning 20th century warming, may have been biased by underestimation of natural climate variations. The bottom line is that if these variations are indeed proven true, then, yes, natural climate fluctuations could be a dominant factor in the recent warming. In other words, natural factors could be more important than previously assumed."[41]

Philip Stott, professor emeritus of biogeography at the University of London: "...the myth is starting to implode. ... Serious new research at The Max Planck Institute has indicated that the sun is a far more significant factor..."[42]

Henrik Svensmark, Danish National Space Center: "Our team ... has discovered that the relatively few cosmic rays that reach sea-level play a big part in the everyday weather. They help to make low-level clouds, which largely regulate the Earth's surface temperature. During the 20th Century the influx of cosmic rays decreased and the resulting reduction of cloudiness allowed the world to warm up. ... most of the warming during the 20th Century can be explained by a reduction in low cloud cover."[43]

Jan Veizer, environmental geochemist, Professor Emeritus from University of Ottawa: "At this stage, two scenarios of potential human impact on climate appear feasible: (1) the standard IPCC model ..., and (2) the alternative model that argues for celestial phenomena as the principal climate driver.

... Models and empirical observations are both indispensable tools of science, yet when discrepancies arise, observations should carry greater weight than theory. If so, the multitude of empirical observations favours celestial phenomena as the most important driver of terrestrial climate on most time scales, but time will be the final judge."[44]

[edit] Believe cause of global warming is unknown

Scientists in this section conclude it is too early to ascribe any principal cause to the observed rising temperatures, man-made or natural.

Syun-Ichi Akasofu, retired professor of geophysics and Director of the International Arctic Research Center of the University of Alaska Fairbanks: "[T]he method of study adopted by the International Panel of Climate Change (IPCC) is fundamentally flawed, resulting in a baseless conclusion: *Most of the observed increase in globally averaged temperatures since the mid-20th century is very likely due to the observed increase in anthropogenic greenhouse gas concentrations.* Contrary to this statement ..., there is so far no definitive evidence that 'most' of the present warming is due to the greenhouse effect. ... [The IPCC] should have recognized that the range of observed natural changes should not be ignored, and thus their conclusion should be very tentative. The term 'most' in their conclusion is baseless."[45]

Claude Allègre, geochemist, Institute of Geophysics (Paris): "The increase in the CO_2 content of the atmosphere is an observed fact and mankind is most certainly responsible. In the long term, this increase will without doubt become harmful, but its exact role in the climate is less clear. Various parameters appear more important than CO_2. Consider the water cycle and formation of various types of clouds, and the complex effects of industrial or agricultural dust. Or fluctuations of the intensity of the solar radiation on annual and century scale, which seem better correlated with heating effects than the variations of CO_2 content."[46]

Robert C. Balling, Jr., a professor of geography at Arizona State University: "[I]t is very likely that the recent upward trend [in global surface temperature] is very real and that the upward signal is greater than any noise introduced from uncertainties in the record. However, the general error is most likely to be in the warming direction, with a maximum possible (though unlikely) value of 0.3 °C. ... At this moment in time we know only that: (1) Global surface temperatures have risen in recent decades. (2) Mid-tropospheric temperatures have warmed little over the same period. (3) This difference is not consistent with predictions from numerical climate models."[47]

John Christy, professor of atmospheric science and director of the Earth System Science Center at the University of Alabama in Huntsville, contributor to several IPCC reports: "I'm sure the majority (but not all) of my IPCC colleagues cringe when I say this, but I see neither the developing catastrophe nor the smoking gun proving that human activity is to blame for most of the warming we see. Rather, I see a reliance on climate models (useful but never "proof") and the coincidence that changes in carbon dioxide and global temperatures have loose similarity over time."[48]

Petr Chylek, Space and Remote Sensing Sciences researcher, Los Alamos National Laboratory: "carbon dioxide should not be considered as a dominant force behind the current warming...how much of the [temperature] increase can be ascribed to CO_2, to changes in solar activity, or to the natural variability of climate is uncertain"[49]

William R. Cotton, Professor of Atmospheric Sciences at Colorado State University said in a presentation, "It is an open question if human produced changes in climate are large enough to be detected from the noise of the natural variability of the climate system."[50]

Chris de Freitas, Associate Professor, School of Geography, Geology and Environmental Science, University of Auckland: "There is evidence of global warming. ... But warming does not confirm that carbon dioxide is causing it. Climate is always warming or cooling. There are natural variability theories of warming. To support the argument that carbon dioxide is causing it, the evidence would have to distinguish between human-caused and natural warming. This has not been done."[51]

David Deming, geology professor at the University of Oklahoma: "The amount of climatic warming that has taken place in the past 150 years is poorly constrained, and its cause--human or natural--is unknown. There is no sound scientific basis for predicting future climate change with any degree of certainty. If the climate does warm, it is likely to be beneficial to humanity rather than harmful. In my opinion, it would be foolish to establish national energy policy on the basis of misinformation and irrational hysteria."[52]

Richard Lindzen, Alfred P. Sloan Professor of Atmospheric Science at the Massachusetts Institute of Technology and member of the National Academy of Sciences: "We are quite confident (1) that global mean temperature is about 0.5 °C higher than it was a century ago; (2) that atmospheric levels of CO_2 have risen over the past two centuries; and (3) that CO_2 is a greenhouse gas whose increase is likely to warm the earth (one of many, the most important being water vapor and clouds). But--and I cannot stress this enough--we are not in a position to confidently attribute past climate change to CO_2 or to forecast what the climate will be in the future."[53] "[T]here has been no question whatsoever that CO_2 is an infrared absorber (i.e., a greenhouse gas — albeit a minor one), and its increase should theoretically contribute to warming. Indeed, if all else were kept equal, the increase in CO_2 should have led to somewhat more warming than has been observed."[54]

Roy Spencer, principal research scientist, University of Alabama in Huntsville: "We need to find out how much of the warming we are seeing could be due to mankind, because I still maintain we have no idea how much you can attribute to mankind."[55]

[edit] Believe global warming will benefit human society

Scientists in this section conclude that projected rising temperatures and/or increases in atmospheric carbon dioxide will be of little impact or a net positive for human society.

Craig D. Idso, faculty researcher, Office of Climatology, Arizona State University; founder of The Center for the Study of Carbon Dioxide and Global Change: "the rising CO2 content of the air should boost global plant productivity dramatically, enabling humanity to increase food, fiber and timber production and thereby continue to feed, clothe, and provide shelter for their still-increasing numbers...this atmospheric CO_2-derived blessing is as sure as death and taxes."[56]

Sherwood Idso, former research physicist, USDA Water Conservation Laboratory, and adjunct professor, Arizona State University: "[W]arming has been shown to positively impact human health, while atmospheric CO_2 enrichment has been shown to enhance the health-promoting properties of the food we eat, as well as stimulate the production of more of it. ... [W]e have nothing to fear from increasing concentrations of atmospheric CO_2 and global warming."[57]

Patrick Michaels, former state climatologist, University of Virginia: "scientists know quite precisely how much the planet will warm in the foreseeable future, a modest three-quarters of a degree (Celsius), plus or minus a mere quarter-degree...a modest warming is a likely benefit."[58]

Here is but a handful of the 31,000 Climate Sceptics. Scientific consensus? Pull the other one!

QED. Quod erat demonstrandum.

Anthony Bright-Paul

May 3rd 2008.

The Death Blow to Anthropogenic Global Warming

Published by Stephen Wilde June 4, 2008

The influence of the sun has been discounted in the climate models as a contributor to the warming observed between 1975 and 1998. Those who support the theory of anthropogenic global warming (AGW), now known as anthropogenic climate change so that recent cooling can be included in their scenario, always deny that the sun has anything to do with recent global temperature movements.

The reason given is that Total Solar Irradiance (TSI) varied so little over that period that it cannot explain the warming that was observed. I don't yet accept that TSI tells the whole story because it is ill defined and speculative as regards it's representation of all the different ways the sun could affect the Earth via the entire available range of physical processes.

Despite the limitations of TSI as an indicator of solar influence I think there are conclusions we can draw from the records we do have. Oddly, I have not seen them discussed properly anywhere else, especially not by AGW enthusiasts. This chart shows the pattern of solar activity since the Maunder minimum:

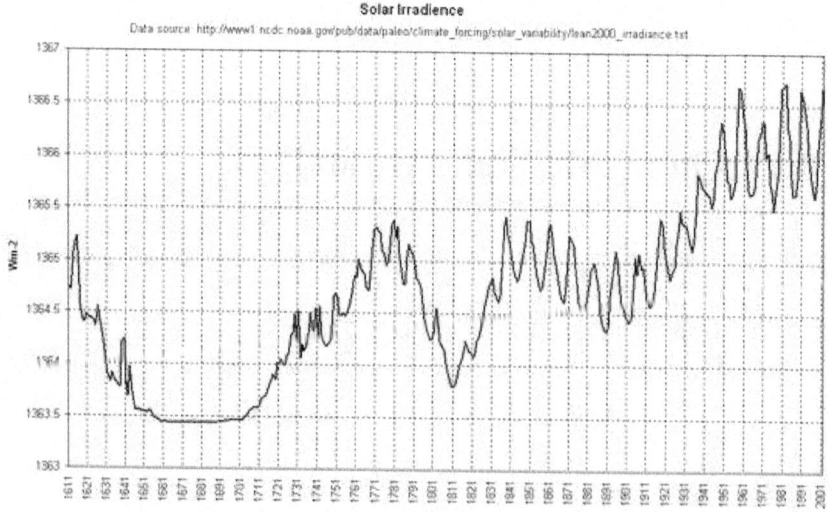

It is true that, as the alarmists say, since 1961 the average level of TSI has been approximately level if one averages out the peaks and troughs from solar cycles 19 through to 23.

However, those solar cycles show substantially higher levels of TSI than have ever previously occurred in the historical record.

Because of the height of the TSI level one cannot simply ignore it as the IPCC and the modellers have done.

The critical issue is that having achieved such high levels of TSI by 1961 the sun was already producing more heat than was required to maintain a stable Earth temperature. On that basis alone the theory of AGW cannot be sustained and should now die.

Throughout the period 1961 to about 2001, there was a steady cumulative net warming effect from the sun. The fact that the TSI was, on average, level during that period is entirely irrelevant and misleading.

It is hardly likely that such a high level of TSI compared to historical levels is going to have no effect at all on global temperature changes and indeed during most of that period there was an enhanced period of positive Pacific Decadal Oscillation that imparted increasing warmth to the atmosphere. My link below to article 1041 contains details of my view that the sun drives the various oceanic oscillations which in turn drive global temperature variations with all other influences including CO2 being minor and often cancelling themselves out leaving the solar/oceanic driver supreme.

It could be said that the increase in TSI from a little over1363 to a little under1367 Watts per square metre over the 400 year period shown is pretty insignificant. However a square metre is a miniscule portion of the surface of the planet so that even a tiny increase or decrease in the heat being received on average over each such tiny area translates into a huge change in total heat budget for the entire planet. The smallness of the apparent range of variation is a function of the smallness of the area subdivision used rather than an indication of insignificance. It is fortunate for us that the sun is not more variable.

The observation of a historically high level of TSI from 1961 to 2001 tends to fit with the theories set out in my other articles about the real cause of recent warming and the real link between solar energy, ocean cycles and global temperatures.

1. Global Warming and Cooling – The Reality

2. The Real Link Between Solar Energy, Ocean Cycles and Global Temperature

Amongst other things the above link (2) shows how the negative PDO from 1961 to 1975 cancelled out the warming effects of solar cycles 18 and 19 and led to a cooling trend during those years despite the relatively high TSI levels. The switch to a positive PDO from 1975 to 2001 allowed the solar warming influence to resume. We now have both a falling TSI and a negative PDO which is an entirely different (indeed opposite) scenario to the one which led to the concerns about runaway warming.

If the current scenario continues for a few more years then real world observations will resolve most of the disputed issues. For the past 10 years the real world has been moving in the direction predicted by the solar driver theory and in my articles I have described the oceanic mechanism that transfers solar input to the atmosphere and then to Space.

If global temperatures were to resume warming despite a reduction in solar activity and/or a negative PDO then the alarmist position might be vindicated. The alarmist camp is predicting such a resumption of warming. The Hadley Centre suggested 2010 but others have more recently suggested 2015. If there is no resumption of warming by 2015 then AGW is dead as a theory. It would not count in favour of AGW if any resumed warming were accompanied by increased solar activity or a positive PDO because that would put the solar driver back in control.

My own view is that there is plenty of evidence currently available that should demonstrate from an objective viewpoint that the theory of AGW is already dead, namely:

1) Real world temperature observations which are diverging from model expectations more and more as time passes

2) The clear recent decline in solar activity

3) The return to a negative (cooling) Pacific Decadal Oscillation) which may last 30 years on past performances

4) A change in global weather patterns which I noticed as long ago as 2000 whereby the jet streams moved back towards the equator from the positions

they adopted during the warming spell. The observation that a global warming or cooling trend can be discerned from seasonal weather patterns seems to be unique to me and will be dealt with in more detail in my next article.

Those who still believe in AGW have to be able to show that any CO_2 driver is powerful enough to seriously disrupt the solar driver. If all that the CO_2 does is to marginally raise global temperature over the period of a natural solar driven warming and cooling cycle then there is nothing to fear because the mitigating effect in cool periods will outweigh any discomfort from the aggravating effect at and around the peak of the warm periods.

In fact, it is possible that even the extra warmth around the natural warm peaks will be entirely beneficial.

There are other interesting implications to be drawn from the TSI history referred to above.

Applying a little logic it must be the case that at a certain level of TSI the global temperature budget will be balanced i.e. neither warming nor cooling. During the 400 years since the world experienced the relatively low TSI levels of the 1600's that point of balance must have been crossed and re crossed many times as the TSI numbers varied with time. That is why the world has experienced warming and cooling spells regularly over the centuries (though with an average warming trend since 1601)

As it happens the chart shown covers TSI from the depths of the Little Ice age to the recent warm spell so it is clear that the point of transition from net cooling to net warming is somewhere within the range 1363 to 1367 Watts per square metre. Indeed on the basis of just a brief glance at the chart that point of transition is obviously lower than the average TSI between 1961 and 2001 hence my assertion that during those years there was a steady solar warming effect which adequately explains the observed warming without reliance on rising CO_2. This is such a simple and obvious point that I really do not understand why the IPCC and the modellers did not see it.

The information that we need and which is critical to the whole global warming debate is some idea of the level of TSI at which the Earth switches from net warming to net cooling. It will be hard to identify because, as I have mentioned in my other articles, the filtering of the solar signal through the various oceanic cycles is neither rapid nor straightforward.

In fact that point of transition will itself vary over time depending on whether, at any given moment, the oceanic cycles are working against or in support of the TSI changes. Similarly the speed of response will vary for the same reasons.

I really do not see how any climate model can operate meaningfully without that fundamental piece of information.

Clearly the 'elephant' is missing from the room.

Finally, in view of the widespread concerns about the involvement of CO2 I should emphasise that if solar energy is the primary driver of global temperature then the only consequence of a stronger greenhouse effect is going to be a slight upward movement of the prevailing temperature throughout the natural warming and cooling cycles.

Because of the logarithmic decline in the greenhouse warming effect of increased amounts of CO2 there is never going to be enough greenhouse effect from any amount of increased CO2 to overturn the primary solar driver or the regular movements from warming to cooling and back again.

The only 'tipping point' we need be concerned with is the level of global temperature at which warming switches to cooling and vice versa. Due to the much greater threat from natural cooling the higher we can lift the global temperature at that tipping point the better. On balance we need more CO2 rather than less.

The band of TSI in which the switch from warming to cooling and back again is a variation of less than 4 Watts per square metre of heat arriving at the Earth's surface.

In view of the size and volatility of the sun we can be boiled or frozen at any time whatever we do. The only reason the sun seems stable enough for us to live with it is that in relation to astronomic timescales our whole existence as a species is but a flash of light in darkness.

The whole of modern civilisation has been made possible by a period of solar stability within a band of less than 4 Watts per square metre. It will not be a result of anything we do if solar changes suddenly go outside that band. On a balance of probability it is more likely that the TSI will soon drop back from the recent unusual highs but remaining within the band of 4 Watts per square metre. It would need the arrival of the next ice age to go significantly below 1363 but even a reduction down to 1365 from present levels could introduce a dangerous level of cooling depending on where the tipping point currently lies.

A period several decades of reduced solar activity will quickly need more emissions producing activity to SAVE the planet yet nonetheless the populations of most living species will be decimated. At present population levels a repeat of the Little Ice Age a mere 400 years ago will cause mass starvation worldwide. Does anyone really think that the CO2 we produce is effective enough to reduce that risk to zero when we have plenty of astronomic evidence of an imminent reduction in solar activity?

And, moreover, the real world temperature movements are currently a good fit with the solar driver theory both as regards the warming spell, the subsequent stall and the recent turn downwards.

The AGW risk analysis process (if anyone ever bothered with one) is seriously flawed.

http://www.newclimatemodel.com/

Published by Stephen Wilde June 4, 2008

No smoking hot spot

Dr David Evans | *July 18, 2008*

In The Australian

I DEVOTED six years to carbon accounting, building models for the Australian Greenhouse Office. I am the rocket scientist who wrote the carbon accounting model (FullCAM) that measures Australia's compliance with the Kyoto Protocol, in the land use change and forestry sector.

FullCAM models carbon flows in plants, mulch, debris, soils and agricultural products, using inputs such as climate data, plant physiology and satellite data. I've been following the global warming debate closely for years.

When I started that job in 1999 the evidence that carbon emissions caused global warming seemed pretty good: CO2 is a greenhouse gas, the old ice core data, no other suspects.

The evidence was not conclusive, but why wait until we were certain when it appeared we needed to act quickly? Soon government and the scientific community were working together and lots of science research jobs were created. We scientists had political support, the ear of government, big budgets, and we felt fairly important and useful (well, I did anyway). It was great. We were working to save the planet.

But since 1999 new evidence has seriously weakened the case that carbon emissions are the main cause of global warming, and by 2007 the evidence was pretty conclusive that carbon played only a minor role and was not the main cause of the recent global warming. As Lord Keynes famously said, "When the facts change, I change my mind. What do you do, sir?"

There has not been a public debate about the causes of global warming and most of the public and our decision makers are not aware of the most basic salient facts:

1. The greenhouse signature is missing. We have been looking and measuring for years, and cannot find it.

Each possible cause of global warming has a different pattern of where in the planet the warming occurs first and the most. The signature of an increased greenhouse effect is a hot spot about 10km up in the atmosphere over the tropics. We have been measuring the atmosphere for decades using radiosondes: weather balloons with thermometers that radio back the temperature as the balloon ascends through the atmosphere. They show no hot spot. Whatsoever.

If there is no hot spot then an increased greenhouse effect is not the cause of global warming. So we know for sure that carbon emissions are not a significant cause of the global warming. If we had found the greenhouse signature then I would be an alarmist again.

When the signature was found to be missing in 2007 (after the latest IPCC report), alarmists objected that maybe the readings of the radiosonde thermometers might not be accurate and maybe the hot spot was there but had gone undetected. Yet hundreds of radiosondes have given the same answer, so statistically it is not possible that they missed the hot spot.

Recently the alarmists have suggested we ignore the radiosonde thermometers, but instead take the radiosonde wind measurements, apply a theory about wind shear, and run the results through their computers to estimate the temperatures. They then say that the results show that we cannot rule out the presence of a hot spot. If you believe that you'd believe anything.

2. There is no evidence to support the idea that carbon emissions cause significant global warming. None. There is plenty of evidence that global warming has occurred, and theory suggests that carbon emissions should raise temperatures (though by how much is hotly disputed) but there are no observations by anyone that implicate carbon emissions as a significant cause of the recent global warming.

3. The satellites that measure the world's temperature all say that the warming trend ended in 2001, and that the temperature has dropped about 0.6C in the past year (to the temperature of 1980). Land-based

temperature readings are corrupted by the "urban heat island" effect: urban areas encroaching on thermometer stations warm the microclimate around the thermometer, due to vegetation changes, concrete, cars, houses. Satellite data is the only temperature data we can trust, but it only goes back to 1979. NASA reports only land-based data, and reports a modest warming trend and recent cooling. The other three global temperature records use a mix of satellite and land measurements, or satellite only, and they all show no warming since 2001 and a recent cooling.

4. The new ice cores show that in the past six global warmings over the past half a million years, the temperature rises occurred on average 800 years before the accompanying rise in atmospheric carbon. Which says something important about which was cause and which was effect.

None of these points are controversial. The alarmist scientists agree with them, though they would dispute their relevance.

The last point was known and past dispute by 2003, yet Al Gore made his movie in 2005 and presented the ice cores as the sole reason for believing that carbon emissions cause global warming. In any other political context our cynical and experienced press corps would surely have called this dishonest and widely questioned the politician's assertion.

Until now the global warming debate has merely been an academic matter of little interest. Now that it matters, we should debate the causes of global warming.

So far that debate has just consisted of a simple sleight of hand: show evidence of global warming, and while the audience is stunned at the implications, simply assert that it is due to carbon emissions.

In the minds of the audience, the evidence that global warming has occurred becomes conflated with the alleged cause, and the audience hasn't noticed that the cause was merely asserted, not proved.

If there really was any evidence that carbon emissions caused global warming, don't you think we would have heard all about it ad nauseam by now?

The world has spent $50 billion on global warming since 1990, and we have not found any actual evidence that carbon emissions cause global warming. Evidence consists of observations made by someone at some time that supports the idea that carbon emissions cause global warming. Computer models and theoretical calculations are not evidence, they are just theory.

What is going to happen over the next decade as global temperatures continue not to rise? The Labor Government is about to deliberately wreck the economy in order to reduce carbon emissions. If the reasons later turn out to be bogus, the electorate is not going to re-elect a Labor government for a long time. When it comes to light that the carbon scare was known to be bogus in 2008, the ALP is going to be regarded as criminally negligent or ideologically stupid for not having seen through it. And if the Liberals support the general thrust of their actions, they will be seen likewise.

The onus should be on those who want to change things to provide evidence for why the changes are necessary. The Australian public is eventually going to have to be told the evidence anyway, so it might as well be told before wrecking the economy.

Dr David Evans was a consultant to the Australian Greenhouse Office from 1999 to 2005.

Black Magic

Anthony Bright-Paul
February 5th 2009

Never in the history of mankind has Black Magic been practiced more than it is today. And never has man been less aware of this same magic. The practice of suggestion, the suggestibility, indeed the gullibility of man at this present time has never been greater. So we see that not unintelligent people are likewise afflicted – Heads of States, Chancellors, Archbishops, Princes and so on.

Nothing illustrates this better than the current hysteria over Climate Change, when many well-meaning people have clearly taken leave of their senses, persuaded by the Black Magicians.

It so happens that in these days snow has carpeted Britain and more is forecast. In the USA the icy weather has reached even as far south as Florida. In Canada where my own correspondents have told me that the weather is brutal with snow in Toronto several feet thick, (not inches or centimetres, note!) the extreme cold has penetrated even to British Columbia. While in Melbourne at the recent Australian Open the temperatures were the opposite, so hot that they had to close the roof on the Rod Laver Court in order to use air-conditioning.

Nobody yet has claimed to be able to control the Weather. Although the Met Office has a hand at forecasting what will happen in the next 24 hours, even they, the experts in this matter, even they do not presume to be able to control anything. They cannot make the wind to blow, the sun to shine, the clouds to gather. They cannot make the rain to fall when there is drought, not can they abate it when there is flood.

In the face of the Weather all mankind are equally humble. If anyone suggested that they could control the forces of Great Nature, they would be laughed to scorn.

And yet we have Prime Ministers and Members of Parliament, German Chancellors, Princes, Archbishops who pontificate about Climate Change! Can one imagine anything more absurd? They actually want to 'arrest' Climate! They believe, because someone has made some computer models, that Climate is running out of control, and that they, the great big panjandrums of this world, can somehow stop the forces of Great Nature.

When one thinks about this soberly for one moment, one is forced to see how utterly ludicrous this is. Even a simpleton knows that Climate has been changing for millennia; the great ice ages being followed by far warmer periods than we have recently experienced, followed again by little ice ages, from the last of which we are only just emerging.

What is even more ludicrous is that all this Climate Change nonsense is based on an unproven hypothesis, in a pseudo-scientific area specially chosen to bemuse the common man and to provoke hysteria. This is what I call Black Magic. For, not only have otherwise intelligent and well-meaning men been persuaded that they can take steps to control Climate, but also moreover that they must so do before catastrophe befalls mankind.

And what are the means that these same potentates, what are the means with which these same arrogant people imagine that they can control the great forces of Nature? Pause for a moment and think about this.

While we have no control of the winds, nor of the ocean currents, while we have no control of the way the Earth wobbles on its axis (the Milankovitch effect), while we have no control of the giant storms on the surface of the sun, while we have no control whatsoever of volcanoes and the movement of the tectonic plates, nor of the mantle within the earth's crust, some bright spark has come up with the idea that we can control Climate, no less, by regulating the emissions of a very minor Greenhouse Gas, Carbon Dioxide.

In spite of the fact that all humans, nay all animals breathe in Oxygen and exhale CO2 (which is done liberally on the floor of the Commons), in spite of the fact that these same members of Parliaments imbibe CO2 every day in the their Scotch and Ginger Ales, or Coca-Cola if so inclined, this innocent and non-polluting gas has been picked upon as the villain of the piece. One does not have to be a scientist to realise how absolutely absurd this is. Yet politicians are still huffing and puffing persuading each other to sign up to the Kyoto Protocol, which can have no effect whatever on Climate, but which may well ruin the economies of the Western World.

We do not need to go into the scientific arguments that CO2 is constantly on the move, that it rises and falls back into the oceans, that it is less than 0.4% of all Greenhouse Gases and that the anthropogenic part is miniscule, because if we do that we abdicate our own responsibilities, we abdicate our common sense.

The truth is that these Black Magicians with their mumbo-jumbo have persuaded a whole mass of people of something that is clearly untenable. Indeed a climate of hysteria has been evoked. It is time now for people, for the common man, to wake up and reject for ever the pseudo-scientific pap with which we are being fed.

It is time to reject the spell of the Black Magicians.

Anthony Bright-Paul

February 5th 2009

Models of Illusion

John Droz, jr.

Physicist and Environmental Advocate

22/07/2009 rev a

Everyone readily admits that things aren't always what they seem. But are we really applying this knowledge in our daily dealings — are we consciously ferreting out the *illusionary* from the *reality*? I think not.

For instance, although the overwhelming evidence indicates that our important policies are being dictated by self-serving lobbyists — that's not the whole picture.

There is another extremely powerful (but much less visible) agent employed by these parties, that may be their most powerful tool.

The person behind the screen is the computer programmer. And, just like in the *Wizard of OZ*, they do **not** want you to look at this real controller.

I'll probably have to turn in my membership card, but as a computer programmer (and physicist and environmental activist) I'm here to spill the beans about the Wiz.

The first hint of trouble is spelled out in *Wikipedia's* explanation about computer programmers: "The discipline differs from many other technical professions in that programmers generally do not need to be licensed or pass any standardized (or governmentally regulated) certification tests in order to call themselves 'programmers' or even 'software engineers.'" Hmmm.

My layperson explanation is that computer programming is all about making assumptions, and then converting these into mathematical (Boolean) equations.

The big picture question is this: **is it really possible to accurately convert complex real-world situations into one's and zero's?** Hal may think so, but higher processing brains say no. Yet this is continuously attempted, with very limited success. Let's pull the screen back a bit more.

We'll start with an example about how such a model makes assumptions.

One of the computer programs I wrote was for debt collectors. A typical scenario was that a debtor was given a date to make a payment, and the collection company didn't receive it by that time. *What response is then appropriate?*

In such a circumstance the computer program typically makes an automatic contact with the debtor. (Remember there are thousands of these debtors, and it would be prohibitively time consuming for an agency person to manually check into and follow up each case.)

So what to say in this correspondence to the debtor? Well, it comes down to the assumptions made by the computer programmer.

The programmer tries to simplify such situations into mathematical options. In this case they may decide that the question is "does the debtor have the money to make this payment: YES or NO?" This relatively basic choice then leads to a Boolean progression within the program.

How does the programmer (model) decide on YES or NO? Well other indicators would be used (e.g. were prior payments made on time) to come up with a statistical probability.

Of course any computer model is not ONE set of choices, but rather a whole **series** of YES/NO (IF/OR) calculations that lead to a conclusion. In a complex situation (e.g. debt collection, climate change, or financial derivatives) there could easily be a HUNDRED such choices to deal with.

To understand the implications of that, let's just consider the case where there are TEN such decision points — each with a YES or NO answer. At the end of such a pipeline, that means that there are 2^{10} (i.e. 1024) possible results. **That's a LOT of different potential conclusions!**

Unfortunately there are actually MANY more possibilities! The assumption that this debtor situation could be condensed down to a YES or NO answer, is not accurate. There are several other real situations that fall outside of YES or NO.

For instance, what if the debtor never got a notice that the amount was due by the date the agency is monitoring? Or what if the debtor sent the money and it got lost in transition? Or what if the debtor made the payment to the original person they owed, rather than the collection agency? Or what if the debtor sent in the money on time, and the collection agency incorrectly didn't credit the debtor for the payment? Etc., etc.

For the computer program (model) to be accurate, ALL of these scenarios need to be able to be handled properly (legally, timely, etc.). Can you begin to see the complexity here, just with this very simple example of a payment not being received on time?

There is still another significant factor (we're up to #4 now) not mentioned yet. What about the situation where the debtor hasn't paid, but it's because his child has ALS, and he has no insurance? How does a computer programmer write code for more abstract concepts, like "fairness"? In other words, can *ones* and *zeros* be arranged in such a way to represent intangibles? I think not.

So the bottom line question is this: *is there any way that a computer program can correctly handle ALL of these real-world possibilities — even in this simple debt collection case*? The answer is no. **NO!!!**

Another perspective is that we have considerable difficulty just translating the relatively simple thing we call language — e.g. Greek biblical texts into English. *How many versions of the Bible are there?* **Why isn't there just one?**

Can we possibly hope to translate a process much more complicated than just words? We can certainly try, but clearly the answer is that there is a LOT lost in the *translation* of any complex scenario (debtors, energy performance, etc.) into mathematical equations and computer code.

Some uninformed parties believe that the user has control of all the variables, and can manually (and accurately) change scenarios. That is incorrect, as the user-controlled elements only represent **a small fraction** of the actual number of factors that are built into the computer model.

A similar fallacy is to think something like "we know the assumptions that the programmers made, and are adjusting accordingly." Wrong!

In writing a computer program of any complexity, there are literally **hundreds** of assumptions made. **The computer programmer does NOT reveal all these to his customer, for much the same reasons that an accountant does not tell his client all of the assumptions made in preparing a tax return.** He goes over a few of the more basic items, and then says "sign here."

Oh, yes, this example brings up still another MAJOR variable (#7): **the data the programmer uses as the basis for his creation.**

Just like preparing a tax return depends on two parties working together, writing a computer model is a collaboration between scientist and programmer. If the taxpayer gives incomplete or inaccurate data to the accountant, the result will be wrong. *What's disconcerting is that in many cases, neither party will know that the results are in error...*

Similarly if the scientist gives incomplete or inaccurate date to the programmer to use in his creation, the result will likewise be wrong. **AND neither party will know it!**

I hate to keep going on here, but this is important stuff! Believe it or not, there is still one more significant variable (#8) that we have to take into account. After a computer model is generated, there is then an **interpreter** (e.g. IPCC) that translates the "results" for politicians and the public (i.e. the media).

Here's a surprise: these public interpretations are influenced by such factors as political, religious, environmental, financial, and scientific *opinions*. In their public revelations, do the interpreters explain all of their underlying biases? By now you know the answer: **absolutely not.**

When these are introduced into the equation we obviously have strayed so far from scientific fact that it is not even in sight anymore.

Soooo, we need to think VERY CAREFULLY before we take major actions (e.g. spend a **Trillions** of dollars based on climate predictions, wind energy projected performance, etc.) that are almost entirely based on computer models.

What to do? Should we just scrap all computer models?

No, that's the other extreme. Computer models have merit — *but shouldn't be the tail wagging the dog.*

We should realistically see computer models for what they are — tools to assist us in organizing our thoughts, and producers of highly subjective results that are simply starting points for real scientific analysis.

Because of their inherent limitations (which I've just touched on here) ALL computer models should be treated with a very healthy degree of skepticism.

To insure appropriate integrity, ALL computer models regarding matters of importance should be subjected to the rigors of **scientific methodology.**

*If they can't **accurately** and **continuously** replicate the results of real world data, then they should be discarded.* Unfortunately that is not what is happening.

We have gotten so addicted to the illusion that these programs are accurate — and some have become so agenda driven — that we are now adjusting or discarding real world date that doesn't agree with the model. **This is insane!**

If a model has not been proven to fully reflect reality, then it has very limited use, and should be treated with the same degree of consideration that one might give a horoscope.

John Droz, jr.

Physicist and Environmental Advocate

7/22/09 rev a

Singing from the same Hymn Sheet

Anthony Bright-Paul
20.08.2009

William Happer: The Truth About Greenhouse Gases

New GWPF Briefing Paper

London, 17 August - The Global Warming Policy Foundation today published an outstanding briefing paper by the distinguished physicist Professor William Happer of Princeton University (USA).

In his paper *The Truth About Greenhouse Gases*, Professor Happer criticises the misguided scare-mongering about CO2 emissions as well as the habitual exaggeration of the likely impact and risks posed by global warming. He particularly laments the co-option of climate science by governments.

Happer discusses what he calls the "contemporary moral epidemic" of climate alarmism: the notion that increasing atmospheric concentrations of greenhouse gases, notably carbon dioxide, will have disastrous consequences for mankind and for the planet and advocates a sober and balanced assessment based on empirical observations, not computer models.

"CO2 does indeed cause some warming of our planet. Other things being equal, more CO2 will cause more warming. The question is how much warming, and whether the increased CO2 and the warming it causes will be good or bad for the planet," Happer writes.

William Happer is the Cyrus Fogg Brackett Professor of Physics at Princeton University. He is a member of the GWPF's Academic Advisory Council.

The Truth About Greenhouse Gases

Now, Sceptics and Sceptic Professors, does is not make your hearts feel good to see that yet another evidently eminent Professor of Physics has come out boldly on behalf of the sceptic cause? But is there anything in his statement that gives you pause?

I am no Professor, I have no degree of any kind, I am personally only concerned with logic. I am concerned with Facts and Conclusions. In reading the History of Science what one sees is that it takes time for Facts to be established and even longer for the correct Conclusions to be arrived at. This process is ongoing.

There is one thing however that does concern me about Professor Happer's statement, which is, "CO2 does indeed **cause** some warming of our planet." Why does that worry me? It is because that is precisely the argument that the Alarmists use. Please note the use of the word 'cause'. This means that Professor Happer considers that Carbon Dioxide can be a Driver of Climate, which is precisely the argument that Al Gore puts forward and that most sceptics declare is false. Professor Happer goes on "Other things being equal, more CO2 will **cause** more warming."

Once again he asserts that CO2 is causative, if only in a small degree. But is that logical? What is the basis for claiming that Carbon Dioxide can cause warming? Lord Monckton takes the same attitude, proclaiming that the Greenhouse Effect is real, based on the easily replicable Tyndall experiment. But what does the Tyndall experiment show? That Carbon Dioxide absorbs strongly in the infrared and the near infrared. Nobody will quarrel with that, as an established Fact.

But what conclusion can be derived from that? Since Carbon Dioxide absorbs heat, does that infer that it causes heating? Sorry, no way is that logical. For we also know that Carbon Dioxide can also be cooled to make Dry Ice. Ergo, Carbon Dioxide is reactive and not pro-active, not an agent, not a Driver.

(Just imagine for one moment that the Sun stopped shining, that the radiation from the Sun ceased, that there were no TSI, no Total Solar Irradiance, would Carbon Dioxide keep us warm one instant?)

This is even clearer with Water Vapour. We all can experience humidity on a hot day. Does the humidity create the heat? Obviously not, because we know that freezing fog is also Water Vapour. In both cases the gas is passive. In simple language they both can be cooled and be heated. There is no way that they can cause heat.

Yesterday at midday here in England, in Farnborough where I live, it was 22°C and very humid, so much that I was reluctant to get up and mow the grass, sitting by my table in the garden. This varnished tabletop was noticeably hot to the touch. Did I conclude from that, that the tabletop was heating the garden? No way, for it was clear that the day before when we were inundated with rain the table was likewise cool.

Now did the humidity, the water Vapour make the temperature hotter, or just more uncomfortable? Or could it even have made it cooler? Here I will quote from Ian Plimer's book Heaven + Earth, p433.

"Water vapour is an amplifier rather than a trigger. This can easily be put to the test. In a humid coastal area with a clear sky it is warm in the day and mild at night. At the same latitude inland where the air is dry, it will be hot in the day and cold at night."

Here is an interesting use of the word 'trigger'. So Water Vapour is not a cause. Had there been no Water Vapour then the temperature would likely have been higher and the evening colder. This seems to bear out the contention of Hans Schreuder that far from warming, the so-called Greenhouse Gases actually have a cooling effect.

Al Gore is clearly unable to distinguish cause and effect – let that not be said of any sceptics.

Professor Happer '... advocates a sober and balanced assessment based on empirical observations, not computer models.'

There we must all agree with him. Based on empirical observations! Absolutely. But, is there one shred of evidence that man has in any way, shape or form been able to cause warming? Is there any one piece of evidence that shows that any warming of the atmosphere can be attributed to man that cannot be simply attributed to Great Nature?

So Alarmists will cobble together disparate observations that Arctic Ice is melting, even that the North West Passage is open. Such seasonal melting can easily be explained by warming ocean currents. The calving of Arctic glaciers, as Jim Peden, the world-renowned Astrophysicist pointed out, is a sign of advancing, not retreating glaciers. As far as I am aware there is no empirical observation whatsoever that proves conclusively that man has caused any warming whatsoever. Not even Professor Karoly could produce a summer's day.

Svante Arrhenius wanted to make the world warmer – who can blame him? – but all the evidence is that everything that is hot always cools. Is there any one single thing that grows hotter and hotter by itself?

Here simply on the basis of Philosophy and Logic I hope and pray that all Sceptics will get together and sing from the same hymn sheet. Certainly I would agree with Stephen Wilde that the very last thing we would want is for Sceptics to attack each other. Therefore if my Facts are wrong, I am willing to be corrected. Likewise if my reasoning, if my logic is flawed, I am ready to be corrected. Nevertheless, if we are to say that CO_2 **causes** warming, **causes** Global Warming in any way, shape or form, we are creating a hostage to fortune; we are ceding one of the principle bulwarks of scepticism to the Alarmists.

Given the Adiabatic Lapse Rate, given that the atmosphere above approx 7,500 feet is Zero°C and declining, and is about 45°C minus at 30,000 feet, how can heat be reflected down from that altitude where no heat exists? Let Professor Trenberth make diagrams that make no sense, and which according to the Rev Philip Foster do not even add up, but don't let Sceptics get caught up in such confusion. There cannot logically be greenhouse gases except in a greenhouse. It is easy enough to trap a butterfly with a fine net, but try to trap it with air, with any gas or combination of gases, that is impossible. So gases may be heated, may even be burned, may be compressed, may even be liquefied, but in no way can a gas trap anything. The amount or the volume of the gas is entirely irrelevant.

That Greenhouse Gases insulate is another matter. Insulation works two ways. It retards the entrance of heat, as it likewise retards the exit of heat. But in no way does insulation create heat.

It is vitally important that the Sceptics agree on this fundamental, so that it is entirely logical. It must be simple and logical, so that the Sceptic cause is presented in a simple and logical way to the average man in the street. It is clearly totally illogical to suggest that any gas whatsoever drives Climate. Physics deals with the mechanics of how things happen. But let us never confuse the mechanics with the cause, for that is what the Alarmists do.

.

Anthony Bright-Paul
August 20th 2011

Postscript: Here is an interesting excerpt from John Gribbin's "The History of Science."

Joseph Priestley began his experiments involving 'airs' during his time in Leeds, where he lived close to a brewery. The air immediately above the surface of the brew fermenting in the vats had recently been identified as Black's 'fixed air' and Priestley saw that he had a ready-made laboratory in which he could experiment with large quantities of this gas.

He found that it formed a layer roughly nine inches to a foot deep above the fermenting liquor, and that, though a burning candle placed in this layer was extinguished, the smoke stayed there. By adding smoke to the Carbon Dioxide layer, Priestley made it visible, so that waves on its surface (the boundary between carbon dioxide and ordinary air) could be observed, **and it could be seen flowing over the sides and falling to the floor.**

From this we know that carbon dioxide is one and a half times heavier than air, and must therefore be subject to the laws of gravity. In which case it can only be carried upwards by convection, and must fall earthwards subsequently, like cinders from a bonfire.

Emails from James Peden, Astrophysicist.

21.02.2009

Dear Anthony -

Good luck trying to convert the religiously fervent followers of Al Gore and Jim Hansen. It has been my observation that most of the AGW hysterians are not capable of understanding the actual science when it is presented to them, and many refuse to accept it even when it has been clearly delineated.

The fact remains that there is no mechanism at work in the atmosphere, which acts as a "greenhouse"... and physically "traps" heat so it cannot escape. "Greenhouse Gases" (very poorly named) only serve to delay cooling a bit, and without them the planet would be about 34 degrees cooler than it is.

And, most importantly, CO_2 is a very minor player in all this, with water vapour being the principal "thermostat" that regulates our atmospheric temperatures. Have your friends read THIS ... then ask them what they find particularly objectionable. Their mute silence will reflect their academic qualifications to discuss the subject. Remember, the "Warmers" rely on failed computer models.... not actual climate science... to promote this massive hoax.

Cheers,

Jim Peden

Call me Jim, Tony...

Indeed, CO_2 is colourless, so why do the AGW fanatics show smokestacks of dirty dark stuff? To mislead the public.

A Carbon Monoxide molecule does indeed want to combine with another atomic Oxygen atom to form a Carbon Monoxide molecule (which is harmless), but until it does, Carbon Monoxide is quite toxic to man. The rapidity with which this occurs on the availability of free oxygen, obviously. None of us who claim the AGW scare is a giant hoax deny that "pollution" is not good and should be combated as strongly as possible. To refer to CO_2 as a "pollutant" is strictly the product of a low I.Q.

Tim Ball and Bob Carter are both very sharp folks, both of whom often copy me on various climate opinions we tend to swap back and forth here in the "denier" community. I find most of the AGW "realists" to be very bright, well educated, knowledgeable, and often older and/or retired - thus not in danger of losing their jobs (as are many of the younger dissenting scientists, who can't speak out very loudly). I fall in this category.... being able to comment freely because I have no professional mule in the race at this time.

Even if you are "not a scientist" you can read plain language summaries and understand the basics... provided you know which summaries to trust. None of us are 100% correct on any of this... including myself. Whenever one tries to simplify something so the layman understands it, he naturally treads on accuracy and completeness. You can't have it both ways. (Did you read the PDF at the link I sent you?) For 99% of the population, it is Greek and relatively meaningless. But it is highly significant, for all the argument and debate in the world can't change the laws of physics.

Cheers again,

Jim

09.06.2010
When did we transition from scientific analysts to activists? My own personal discovery of the Great Global Warming Hoax came only after taking a closer look at the atmospheric physics involved... having previously assumed that science had done it's job properly and there was a general consensus at the time that we humans were crapping in our nest again. I didn't take long to learn that nothing in the climate of today or in the recent past lay outside of the normal variability of natural climate change, and for whatever minor role it played in "global warming", the 14.77 micron band of CO_2 was

pretty well saturated so adding more at this time posed little risk for the climate and much benefit for agriculture and humans as a result.

For me, it should have ended there, just sit back and chuckle while watching the rest of the (scientifically illiterate) world worship the hockey stick without knowing it actually represented one of the most disastrous mathematical blunders of all time. A good laugh, no more... sort of like the Piltdown Man.

I think it was when the alarmists started talking "carbon taxes" in the trillions of dollars did I realize the entire alarmist community represented a clear and present danger to the future financial security of the United States. It also gave me some insight into the National Institutes of Health statistic that 26% of our population is not mentally wired right... which explains otherwise intelligent people believing in magic energy amplifiers in the form of "forcings" which craps all over the 2nd Law of Thermodynamics but clearly proves that New York is going to be under water soon.

Yes, I suppose I am now an activist, trying to salvage a bit of the financial future for my daughter, if not for me. What I'm seeing here is the end of defence and the beginning of an activist offense from the "skeptics". Count me in, I still see a future worth saving....

Jim Peden

As far as physics of the upper atmosphere was concerned I was extremely lucky to be teamed with some of the best in the business in my day, including Wade Fite and Ted Brackmann at Pitt, who were pretty much some of the pioneers in our particular brand of molecular beam work. Papers were published in highly distinguished journals such as the Journal of Chemical Physics, and the resulting data eventually incorporated into the AstroChemistry Database. After your work becomes the Gold Standard for those particular reactions, you don't have a lot more to prove to anyone else.

I also was involved in some highly classified research, some of which I note has apparently only recently become declassified. We developed a method of detecting over-the-horizon nuclear events by examining excitation of the upper atmosphere. The spooks used our gear to monitor the earliest Chinese nuclear tests.

I was involved in atmospheric physics for only a few years, then I developed an interest in aero systems engineering, where I became a project manager and general technical bureaucrat, with about 20 other engineers working for me as pioneers in digital cockpit instrumentation. I am now fully retired from physical science research, and am, as previously stated, no more than a pundit and thorn in the side of the AGW fraternity. I also dabble in theoretical mathematics, and after a number of years working on my own, managed to come up with an interesting proof of the elusive 4-color theorem in topology that at the time had never been proven.

Cheers,
01.10.2011
Jim

I'm really getting sick of the term "greenhouse effect". We all know there is no process at work in the atmosphere that even remotely resembles a classical mechanical greenhouse.... an invisible, impenetrable boundary layer above which trapped gasses can not pass and can no longer come in thermal contact with the rest of the atmosphere above. One is led to believe a 747 can fly through this "greenhouse roof" but an ordinary air molecule cannot.

Yes, there is a "blanket effect" which only regulates how long it takes to cool down after the sun sets, but the principal blanket "material" is water vapour, not CO_2. A dry desert cools almost immediately after the sun sets; and steamy moist rain forest does not. The difference is water vapour, and the CO_2 content of both regions is approximately the same.

Why do we continue to allow the AGW morons to use that term without challenging them at every turn? The general public cares not about radiation or anything else of a technical nature. They only respond to oft-repeated sound bites. "Repeat a lie often enough, and it becomes the truth....".

Except, of course, it never does.....

Jim P

04.12.2011

Anthony, there seems to be some confusion here, perhaps a semantic error.

"Temperature" is based on a measure of the energy of molecular motion... and indeed, the temperature at the edge of our atmosphere is quite "hot" ... because the molecules, albeit few in number, have a high kinetic energy ... thus technically have a high "temperature".

However, there are very few of them. Therefore the "heat content" is very small.... resulting in very few calories per unit volume. At sea level, there is a pretty good correlation between temperature and heat content: a kettle of boiling water has both a high temperature and high heat content.

But at the edge of space, with very few molecules per unit volume, you have the seemingly paradoxical condition of both high temperature and low heat content.

Ordinary thermometers work by transfer of heat energy from the surroundings to the thermometer. At the edge of space, they simply don't work because there aren't enough surrounding air molecules to counter the natural cooling of an object by radiation. So, trying to measure the temperature via normal methods results in an erroneously low reading. We must remember that all bodies emit Infrared radiation and thus "cool" in the process. A thermometer may read quite low at very high altitudes not because the surroundings are "cold" but because the thermometer is losing heat by radiation and there aren't enough surrounding "hot" air molecules to counter that cooling.

At the Kármán line ... the so-called "edge of space" (about 100 km) there is in fact an abrupt rise in temperature... as solar radiation reacts with the few molecules still in that region, increasing their thermal energy, and thus raising their "temperature".

By coincidence, some of my earliest work in atmospheric physics was on ion-molecule reactions the upper atmosphere, so I have perhaps a bit more knowledge on that particular zone than the average physicist, just sayin.... In fact, one of the "trick" questions I used to give my students was, "How cold is it at the upper edges of our atmosphere?". Most of them would naturally say, "very cold" on first consideration.

And I'm still a strong "denier"... it's just that this particular argument doesn't work for me. Below is a graph of temperature vs altitude, showing the point where the atmosphere stops "cooling" and begins to get "hotter"... at the so-called Kármán line previously mentioned.

Cheers,

Jim Peden

================

The inconvenient truth about ozone-puncturing Two Jags

By Richard Littlejohn

Daily Mail
Last updated at 8:32 AM on 04th September 2009

The preposterous figure of Two Jags has apparently been reincarnated as something called the Council of Europe's 'rapporteur' on climate change. That's a new one on me. Wasn't the dwarf in Time Bandits called Rapporteur?

I've no idea what a rapporteur does, but I would imagine it involves a lot of first-class air travel, five-star hotels and lobster suppers. There's probably a bird thrown in, too.

Two Jags is flying to China this week to deliver a lecture on global warming. That's right, he's jetting halfway round the world and back to talk about the need to cut carbon emissions.

Don't these people have any idea how ridiculous they are?

What astonishes me is that anyone, especially in my trade, takes him seriously. Two Jags is a circus act. Come to think of it, the dwarf in Time Bandits had considerably more gravitas than Prescott.

Yet, in some quarters, he's treated as a proper person. Yesterday's Independent carried an interview with Two Jags, in which he announced that Europe's target of cutting emissions by 80 per cent was nowhere near tough enough. The paper even ran an editorial praising Prescott's authenticity.

The only thing authentic about this old fraud is his ocean-going, ozone-puncturing hypocrisy and self-importance.

I didn't christen him Two Jags without reason. This was the man who insisted on having not one, but two 'gas guzzling' limousines.

He had his wife chauffeured 200 yards along the seafront at the Labour conference so her hairdo wouldn't get windswept. Pauline's creosote-thick hairspray has probably done more damage to the atmosphere than a fleet of SUVs.

Two Jags took a helicopter back to central London from the rugby league final at Wembley and commandeered an RAF flight to turn on the Blackpool illuminations.

After giving a speech on the importance of public transport to the railwaymen's union in Scarborough, he made an ostentatious display of boarding a train home.

He then got off at the next station, where his driver was waiting with the Jag to convey him back to London in air-conditioned, eight-cylinder, 15-miles-to-the-gallon luxury.

Once he returns from China, he will embark on a tour of Britain, lecturing schoolchildren about global warming. We can assume he won't be travelling by bike.

When he was 'in charge' of the environment, he was so concerned about the delicate eco-balance that he ordered tens of thousands of houses to be built on flood plains.

Yet now we are asked to believe that he is a born-again Al Gore. According to the Indy, he is the brilliant global player who brokered the Kyoto deal in 1997 and 'is now returning to a major role in climate change politics'.

All you need to know about the Kyoto 'deal' is that the rest of the world ignored it, while here in Britain it has been used as a catch-all excuse for everything from the extortionate tax on petrol to fining people £500 for putting out their dustbins on the wrong day.

In one sense, I suppose you could argue that Kyoto was a success, since the world has actually been getting colder over the past decade, despite China opening a new coal-fired power station every five minutes.

That inconvenient truth has not deterred the climate-change industry from cranking up the rhetoric, inventing ever-tougher targets and dreaming up an exciting range of new rules, fines and punishments.

Britain's ridiculous obsession with 'man-made global warming' has prevented our building a new generation of power stations.

As a result, we are facing the looming prospect of rolling power cuts in the not-too-distant future.

But you won't find a forest of windmills in the back garden of Two Jags' turreted mansion.

Just as Al Gore consumes enough electricity to power a small town and flies by private jet to deliver his lavishly rewarded pieties on polar bears, so Two Jags, too, thinks that cutting your carbon footprint is for the little people.

This week, The Guardian - which is The Independent with adverts - carried a spread about everyday people who were doing their bit for the planet.

They boasted about how they were going to eat more root vegetables, wear thicker undies and travel by train not car.

A more self-righteous, self-flagellating bunch you'd be hard-pressed to find outside of, er, the pages of The Guardian.

But even though I think they're all barking mad, at least they are prepared to make some kind of self- sacrifice in pursuit of their quasi-religious crusade.

Two Jags was conspicuous by his absence. While those poor, deluded saps are turning down the thermostat, shivering in their thermals and eating their own toenail clippings, you won't catch him chowing down on turnips or taking a slow boat to China.

Our esteemed 'rapporteur' will relax in the rear seat of his limo or at the front of the plane, tucking into the finest food flown in from around the world. And to hell with the ice caps.

He will continue to leave a trail of yeti-sized carbon footprints as he tours the globe lecturing the rest of us on how we're all responsible for razing the rainforests.

In the great debate about nonexistent global warming, this freeloading, flatulent, frequent-flying fool is about as relevant as the polar bear on Fox's glacier mints.

Climate Change – what do we know?

Anthony Bright-Paul
September 26th '09

Without going back into pre-history, without going back hundreds of thousands of years to that sort of time that only specialists, Paleo-Climatologists, can deal in, what do we know about Climate Change in relatively recent times?

Certainly we know that there have been warm periods and cold periods. I am not even going to use the words 'glacials' and 'inter-glacials' as such terms are unfamiliar to the common man. But even a person with a modicum of education knows that there have been great ice ages and wonderful warm periods.

So I will start with the Roman Warming from 250BC to 450AD. This was indeed a remarkable warm period and we can see the evidence today in the magnificent buildings by both Greeks and Romans. We can see the Coliseum in Rome, the amphitheatres at Arles and Nimes and the amazing aqueduct, the Pont du Gard, in the south of France. In Athens we can see the Parthenon, and doubtless there are many other remains scattered all around the Mediterranean that showed the great advances of civilisation during this warm period, that was at least 2° higher than today.

This was followed by the Dark Ages, a bitterly cold period of crop failure, famine, disease, war, depopulation and expansion of ice.

In turn the Mediaeval Warm Period followed this from 900 – 1300AD. At this time the Vikings populated Greenland, which was some 6° warmer than today. In Europe it was a time of building of monasteries, universities and the great Cathedrals.

Then occurred the Little Ice Age with a decrease in solar activity. It was desperately cold, with crop failures, famine and the plague. The Vikings on Greenland died out.

From 1850 onwards till 1998 the Earth grew warmer, as we all emerged from the Little Ice Age. Even now we are still colder than the Roman or Mediaeval Warm Periods, let alone the Holocene Maximum that goes farther back.

From 1998 to the present day there has been stasis or cooling, as per the Hadley Centre part of the Met office, and further cooling is projected for the next 20 years, as Global Warming suffers a blip.

So what can we deduce from this short potted history of recent Climate? One thing is clear beyond peradventure that Climate is always changing. It always has changed and it always will. We can no more stop Climate Change than we can stop the movement of the tectonic plates, or the eruption of submarine volcanoes. We know that in the course of time the South of England will sink, while Scotland will rise; we know that the Gaeranger Fjord in Norway will have a tsunami as the sides collapse. We know or can guess that a large part of California will break off at the San Andreas Fault and float off into the Pacific Ocean. We know that volcanoes will erupt erratically and the dust will obscure the sun and lead to cooling. And as to the sun, that great source of heat and light, we know that man cannot control for one moment the solar winds or the presence or absence of sunspots.

Furthermore we can see that warm periods are especially beneficial for mankind, whereas intense cold periods are greatly to be feared.

Immense forces of Nature are at work, which we humans with all our Science but dimly understand. In spite of this politicians all round the world have latched on to an unproven hypothesis that a very minor gas, Carbon Dioxide, can actually trap heat and cause Global Warming. In view of the immense forces of Nature it is truly amazing that such a hypothesis could have gained almost universal credence. It is even more strange when one looks at the universally agreed composition of the atmosphere: -

        78% Nitrogen
        21% Oxygen
        0.95% Water Vapour
        0.0385% Carbon Dioxide

When one arranges the figures in this way one can see easily how small a part is played by Carbon Dioxide, which for some reason has been picked upon as the villain of the piece – which is totally ludicrous, as CO2 is part of the food chain. Without CO2 plants would not grow, and without the process of photosynthesis that plant life carries out, we humans and all the animal kingdom would not have the oxygen with which to breathe.

Where does Carbon Dioxide come from? In the first place some 40% comes from all the animals, which includes us humans, just by the act of exhaling. We breathe in oxygen and we exhale Carbon Dioxide. Some 57% comes from the oceans and the earth. As the sun shines on the waters streams of CO2 enter the atmosphere. Likewise there is out gassing from the earth and rocks. So we can see that 97% of carbon dioxide is purely of natural origin. **That leaves only 3% that is man-made by the combustion of fossil fuels**.

So then let us take 3% of 0.0385% (sometime called 385 parts per million) and what do we arrive at? The anthropogenic portion of CO2 is 0.001155% of the atmosphere. Built around these figures an alarmist edifice has been created that completely ignores two facts. The first is that in the afore-mentioned potted history, the temperatures have been much higher than those of today. The second is that CO2 in the atmosphere in times past has been sometimes ten, nay twenty-five times greater than today. We ignore the history of our planet and the history of climate change at our peril.

But let us look at those figures once again. If, as the alarmists would have it, that global warming is posing a great danger and that we are facing unprecedented climate change, and that this has been caused by a recent increase in carbon dioxide emissions, then what is their solution? It is quite simply to cut the man-made, anthropogenic, emissions of CO2.

Let us suppose that they are right for a moment. Not even Kevin Rudd or Gordon Brown would imagine that they could blanket the oceans. Not even Al Gore or Senator Kerry or even President Obama would seriously suggest that to save the planet half the population of the earth's animal kingdom, that includes us humans, should be eliminated. No, so what is left?

Let us suppose then by universal edict that the whole world industry was shut down, with no more coal-fired furnaces, no motor cars, no aeroplanes, no electricity and no laptop computers – just suppose that that could happen and that all and every source of man-made carbon dioxide emissions were cut off, and we all returned to simply cultivating our back gardens for food, just what would be the result?

The maths are quite simple: subtract 0.001155 from 0.0385 and we arrive at 0.0373. What does that mean? If we put this in parts per million it means a reduction from 385ppmv to 373ppmv. It means hardly any difference at all!! It means quite simply that enormous efforts of time and money are being spent on trying to eliminate a trace of a trace! The fact is that Great Nature produces Carbon Dioxide in an abundance that cannot be matched by man. So that if we were to sacrifice everything that we have gained in manufacturing and technology it would make not one jot of difference.

No wonder that the vast majority of the world's scientists rejected the Kyoto Protocol - some 17,000 initial signatories of the Oregon Petition, which I understand is now up to 23,000. Please note that this is a far greater number than that UNO's politically controlled and motivated IPCC!

How is it then that so many people have been persuaded that something must be done to prevent Climate Change? It is very simple really, if one can understand the meaning and the role of the Pharaohs. We have today in this world a number of 'Pharaohs'. Typically they are politicians, who are skilled at suggestion and at manipulating people. They are skilled at mass propaganda and will seize ruthlessly the means of mass communication, and deny it to those who dissent. Democracy is an illusion fostered by these demagogues.

Is it not typical that the failed Presidential candidate, Al Gore, is in the van of them, aided and abetted by another failed candidate, Senator Kerry? They are not scientists, they are power hungry

Pharaohs. With respect, or as much respect as I can muster, the same goes for Gordon Brown, who is charging off to Copenhagen; to Kevin Rudd who won a landslide victory promising to commit Australia to the Kyoto Protocol, irrespective of whether or not it would ruin the Australian economy. As for Berlusconi his control of the media is legendary. And so we can go on.

All over we have these power hungry moguls, who have little or no science, who have suggested to otherwise intelligent people that they must tackle climate change. Can anything be more risible, more patently absurd than these miniature demagogues strutting the world stage and playing God?

What is even more ridiculous is that the Skeptics are accused of 'denying Climate Change'! On the contrary Skeptics document climate change and know that climate change is inevitable. To quote Professor Ian Plimer, to whom I am indebted for his book "Heaven+Earth", 'that's what climate does'.

Anthony Bright-Paul
September 26th '09

Apocalypse Now in 50 Days

Anthony Bright-Paul
21st October 2009

On October 20th 2009 Prime Minister Gordon Brown announced that the free world had 50 days in which to prevent climate change.

The UK faces a "catastrophe" of floods, droughts and killer heatwaves if world leaders fail to agree a deal on climate change, the prime minister has warned.

He certainly has hedged his bets – floods and droughts. Then killer heatwaves. Now that is something. It is true that heatwaves do kill off quite a lot of old people, but then so does the extreme cold of even our English winters. Perhaps in the confines of the Houses of Parliament he has not noticed that the weather changes daily.

After the 50 days, which by my calculation ends on Wednesday, December 9th this year 2009, can we make a forecast? Can we say that these powerful world leaders, if they agree, will be successful in preventing climate change? Will Climate by edict from Copenhagen be normal forever after?

That begs the question, 'What is normal weather? What is normal climate? What is the norm for say Jakarta? What is the norm for Copenhagen? What is the norm for Paris? Or for Buenos Aires?'

Since these World leaders meeting on a world stage, since they are determined to stop Global Warming in order to stop Climate Change, would they please define normal unchanging climate first of all?

One thing is for certain; whatever edicts and agreements are made it will still be hot and humid in Indonesia, hot and dry in Morocco, mild and moist in the UK, and desperately cold in Greenland with daily regional variations. Warm currents may melt some Arctic ice, while precipitation will increase Antarctic ice thickness. Great Nature will continue on her wayward way, springing an occasional surprise, a forest fire, a dust storm, a hurricane, a blizzard, a tsunami or a drought.

How many delegates will fly into Copenhagen? 6,000 as in Bali? Now, since these delegates in the main, believe that Carbon Dioxide is now controlling World Climate (is there such a thing?) let us hope that they will all do their best to mitigate their Carbon Footprints. May I make a humble suggestion? Perhaps in the great Conference Hall, as a penance, the power could be shut off for half an hour. It would be interesting to see how these world leaders and the numerous acolytes would cope for just half an hour in Copenhagen without light and heating.

And if perchance, by some foul Skeptic plot, the power failed there altogether for a day, then would those delegates experience for themselves the condition of one third of the world's populations, who are deprived of the benefit of cheap electricity.

We kid ourselves of course. The meeting is not about Climate Change, it never was. Seek the hidden agenda, seek the real objection, is the advice given to every good salesman. The real objective is disruption and chaos, the destruction of the capitalist system. The only thing that is certain to come out of Copenhagen is that the economies of the world will be put in a strait jacket, and the ordinary citizens everywhere will have to suffer from the delusions of megalomaniac world leaders, aided and abetted by the elite corps of professional protesters. Sure, there will be some sincere people among them, some people who really have been taken in by all the Al Gore hogwash, which it is certain he no longer believes himself, as he burns his carbon footprint merrily round the Globe in his executive jet.

The justification for this coming jamboree will be made by increasing hype and scare stories. 'Killer heatwaves' by Gordon Brown is but the first salvo.

Climate Fears
By Piers Corbyn
Thursday, November 26,
Climate Depot
UK Scientist: 'Case for climate fears is blown to smithereens...whole theory should be destroyed and discarded and UN conference should be closed'

'We should end this anti-scientific nonsense now' -- UN's 'Copenhagen jamboree is a scandal and it must be stopped'

UK astrophysicist Piers Corbyn, of the long-range solar forecast group, Weather Action, declared that the ClimateGate revelations have rendered man-made global warming fears false. "The case is blown to smithereens and this whole theory should be destroyed and discarded and Copenhagen conference should be closed," Corbyn said in a contentious on air television exchange with an environmental activist with Russia's WWF. The live TV debate with Corbyn appeared on Moscow's RT TV on November 25, 2009. The RT TV's segment was titled "Heating Cheating." See Full Video of Debate here.

The world is cooling and has been cooling for 7 years and the leading scientists, so-called 'scientists' have been trying to hide that evidence," Corbyn said in reference to hacked emails showing top UN IPCC scientists apparently conspiring to manipulate temperature data and exclude scientific studies from peer-review that they did not agree with.

"We should end this anti-scientific nonsense now," Corbyn said. "The data, real data, over the last one thousand, ten thousand or million years, shows there is no relationship between carbon dioxide and world temperatures or climate extremes. Now we can see that actually the people in charge of data have been fiddling it, and they have been hiding the real decline in world temperatures in an attempt to keep their so called moral high ground," Corbyn told host Bill Dod and Aleksey Kokorin, the Climate Program Coordinator for WWF in Russia.

The upcoming UN global warming summit in Copenhagen is a "complete waste of time," according to Corbyn. "The Copenhagen jamboree is a scandal and it must be stopped," he added. "There is a gigantic bandwagon run by governments who want to control world energy supplies and hold back development in the third world. This thing they are doing now is just the same as they are doing in the banking crisis, it is creating a whole bubble of false values," Corbyn explained.

Corbyn said the ClimateGate revelations further revealed that man-made climate fears are not scientifically valid. "Their claims are false, I repeat, they are false, and this theory they've got is like the titanic and it will crash. I would suggest that honest green campaigners who want to preserve biodiversity should get off this [man-made global warming] bandwagon before it sinks," Corbyn explained.

"Carbon dioxide levels are driven by temperatures, not the other way around. There have been big peaks in CO_2 in past...carbon dioxide is actually a good thing for the world," Corbyn explained. "More CO_2 makes plants and animals more efficient," he added.

Piers Corbyn
Nov 25th 2009

IT'S TIME FOR A CLIMATE "ALL CHANGE".

Express re-titled it - "It's Not The End Of The World"

Johnny Ball
21.12.2009

Ever since writing my TV shows in the 1980's, I have been talking to students, teachers and the general public and enthusing about the amazing possibilities for science and technology in the future. But over thirty years I have seen a terrible change in science education. Roll models like Dalton, Faraday and Curie are hardly ever mentioned and most basic science has disappeared.

Kids are introduced to science as something that is life threatening and deprived of exploration through health and safety. They are being brainwashed into believing that science and technology are crippling the earth and our future, when exactly the opposite is true.

Science education has been turned upside down by worry merchants and it is already costing us dearly in a widespread lack of understanding - it is ignorance that breeds fear and we are raising a generation of scared and scientifically unschooled future adults - It is utter lunacy.

This ignorance goes right to the top and the politics of Climate Change at Copenhagen had lost all sense of rationality. At the end of the Copenhagen fiasco, I heard an IPCC spokesman say that a Copenhagen commitment of $100 billion might keep the global temperature rise by the end of the century, down to 1.5° C, but it won't keep it down by 2°C. That statement on behalf of the IPCC is so devoid of scientific thought that these people must surely be on another planet? I so wish they were.

On Tuesday 15th December at the Bloomsbury Theatre, I was slow handclapped by a section of the audience. They had taken well my Climate Change thoughts, but when I said "The University of East 93

Anglia has been caught cooking the books (scientifically, not financially) for the sake of research funding grants," the reaction started. The plain truth can be upsetting to those who have one set opinion.

My claim that Climate Change is not being caused by man made Atmospheric Carbon Dioxide is not based on one scientific fact but a whole raft of them. Let me explain a couple.

HOW THINGS BURN.

John Dalton's 1803 Atomic Theory, forms the basis of all Chemistry. It is explained with just three elements - Hydrogen, Oxygen and Carbon.

If we burn Methane (Carbon and Hydrogen) with Oxygen, the Methane breaks up. The Carbon joins Oxygen to make CO_2. The Hydrogens join more Oxygen to form H_2O or Water. Nothing is lost or gained, but energy is released - that is the Atomic Theory.

Human energy comes from our food, Carbohydrate (Hydrocarbon and Water) and the Oxygen we breathe. As we burn energy, we break up the hydrocarbons and release H_2O and CO_2 in the same way. All burning, rotting or fermenting produces carbon dioxide and water. So how can we say that one is bad and ignore the other?

In normal air, Water is 60 times more present than CO_2, but in rain storm or monsoon climates the ratio is far greater. Atmospheric CO_2 forms less than 0.04% of the whole atmosphere or one particle in 2500. But man-made CO_2 is at most 4% of that or one particle in every 62,500.

The first IPCC computer models which claimed that not just CO_2, but man made CO_2, was causing global warming, had no input for water at all and were scientifically indefensible. Far from carrying a consensus, many scientists resigned from the IPCC, which still used those names as supporters? (The Global Warming Debate, March 1996 - ISBN 0952773406)

But we all know that water effects climate? On a summer's day, with cloud it is mild; no cloud and it is hot. On a winter's night with cloud, temperatures stay mild; with no cloud, the heat escapes upwards and we have morning frost. This is not rocket science - it's as plain as the nose on your face and it cannot be ignored. But the

IPCC ignored it and all the scientists who disagreed with their flawed computer models? On what grounds?

It is estimated that the atmosphere contains 2,750 billion tons of CO_2, which is an enormous amount to frighten people with. But that is still less than 0.04% of the whole atmosphere. Also CO_2 is half as heavy as air again and falls back to earth or is washed out of the atmosphere by rain, all the time.

So atmospheric CO_2 "has to be replaced", to keep a balance - this is part of the essential Carbon Cycle. Volcanoes produce by far the greatest amounts of CO_2. Many are on land, but far more in the oceans, pushing the continents apart at about the rate your finger nails grow. Their CO_2 production is massive.

On a smaller scale, soil releases CO_2 as do plants and animals which produce around 350 billion tons. Ocean life produces perhaps much more - An incredible 80% of all plant life on earth is in the Oceans. However, it is difficult to measure 70% of the entire Earth's surface, so figures are guesses.

Against all that, Industrial man-made CO_2, though growing, is small by comparison; perhaps 24 billion tons or 1/16th plant and animal CO_2 and no more than 4% of replenished CO_2.

CARE OF THE PLANET.

Of course we cannot ignore our impact on the planet and we should not rape the world of all its fossil fuels until they run out. But we have to get our impact into rational proportion and not apply alarmist Man Made CO_2 scare tactics.

But we are lessening our impact commendably. Power Station generators in the past 15 years by getting steam to do more work, now get 64% more energy out of coal, oil, gas or nuclear fuel. Has anyone told you that before?

Rolls Royce, just three years ago, thought they could not further improve aircraft engines unless aircraft became flying wings. But with USA agreement they are producing an engine that spears all aircraft to 7 miles high and then switches to an economy mode for the rest of the journey. Likely saving on fuel? 25%.

Modern cars have doubled fuel efficiency in recent years and are around 90% recyclable and almost 100% reclaimable. I back new

Nuclear Power not because it is carbon free, which it is, but because the new plant will be at least four time more efficient.

However, third world poverty makes their adoption of the latest technologies impossible. We even deprive them of GM crops for idealist reasons, completely forgetting that European soil will grow practically anything but Third World soil is often so delicate that our methods are totally unsuited for them. It is utter selfishness and scientific lunacy.

Copenhagen activists have been asking for CO_2 reductions for the sake of the Third World. But their argument is wrong. They do not need investment to cut energy use and CO_2 production. They need money for more energy to give them a power base, so that their hospitals can function and they can maintain an electrical grid. Then multi national companies can invest and provide jobs and a financial base on which to build towards equality with our disgracefully rich end of the world.

If we scrapped completely the foolhardy and scientifically unsound chase to reduce carbon, while still aiming for greater efficiency, we would have all the money needed to bring the third world out of poverty, save millions of lives year on year and create a fairer and far more balanced world for our children. Now that would be a legacy for our generation to be proud of.

For whom the bell tolls

Anthony Bright-Paul
14.12.2009

For whom the bell tolls it tolls for thee. So yesterday the bells of several Churches and Cathedrals, both Catholic and Protestant rang out 350 times. Why? You may well ask.

Certain of our learned prelates have got it into their heads that Carbon Dioxide must somehow or other be stopped in its tracks at 350 parts per million by volume. Well, I never! I wonder who put that idea into their priestly craniums. What happened to 'I believe in One God, Maker of Heaven and Earth and of all things visible and invisible?' What happened to Great Nature? What happened to the inexorable processes, the checks and balances that Great Nature affords? What has happened to the search for verity? The distrust of malfeasance?

Truly, my Lords, for whom the bell tolls, it tolls for you. Can you imagine that by taking thought you can add one cubit to your stature? Can you truly imagine that by limiting one of the building blocks of life that you can somehow regulate the atmosphere? Have you somehow or other acquired a divine cognisance of how the atmosphere works? Are you really so gullible that you have been taken in by the hysterical outpourings of the unscientific doomsayers?

Now we have Hilary Benn adding in his tuppence worth. We already know that the climate temperatures have been fixed, that much has been read into the tree rings of a desolate bristle cone pine tree. We know now that cooling is really warming according to the hystericals. And now we are told that the sea is turning into acid. So the sea is turning acidic and according to Ed Milliband the sea levels are rising dramatically.

What is the truth, said jesting Pilate, and stayed for no answer. And why did he not stay? And why should we not stay? I will tell you in one – because these politicians are either totally ignorant of the scientific facts, or are deliberately concealing them for their own obscure purposes.

Sea levels are probably rising about 1 inch in every decade. Does that scare you to death? Do you think that by reducing the emissions of smoke in China and India that sea levels will automatically be reduced? The sea contains huge amounts of Carbon Dioxide, and is very slightly alkaline, between pH 7.9 to 8.2. So according to Hilary Benn the sea is turning acidic, and this will affect the fish stocks and the coral reefs. Hilary, Hilary Benn, where did you go to school? Why has the sea not turned into an acid bath long long ago, when the atmospheric content of CO_2 was twenty times higher than now?

The Global Warmers twist and turn. At every twist they are shown to be wrong. So their warnings grow even more dire. We must act now to limit Carbon Dioxide emissions! Really? But what if Great Nature continues to ignore these politicians, who strut the Globe? Please tell me how they are going to limit the CO_2 that Great Nature produces naturally? They know as well as we all know that they cannot do that – even Dr Vicky Pope of the Met Office actually acknowledges that.

The only emissions that can be limited are the man-made ones. So if at Copenhagen the politicians get a deal to limit all anthropogenic emissions of CO_2 to 50% world wide, by 2020, just what effect will that have on Global temperatures? Ask yourselves. Work it out. $0.0385 \times 3\% = 0.001\% \times 50\% = 0.0005\%$.

So let us ask our worldly-wise prelates, would that do the trick? Would that get you and all mankind back to the magic 350ppmv, that you have rung your bells for? Get a Life, please – just do some simple Arithmetic. If the Global Warmers are right in the diagnosis and in the prognoses, note, then the solutions they propose will be absolutely meaningless. They will make no difference at all. No difference that is to Global temperatures, no difference to the seas, no difference to anything except one thing.

You have it in one. **It would sound the death knell for world industry**. For whom would the bells toll then? Oh yes, my friends, the bells would toll for all of us, for *manusia,* for all mankind.

The snake-oil merchants have confused everyone. 40% of the world population have no electricity or running water. That is a real problem. Poverty. Poverty is endemic and the characteristic of the Third World. The next real problem is pollution. How to create energy, electricity, without massive pollution? We all agree that pollution is a huge problem. The leaders in China and India know that. The Western world knows that.

Pollution comes from smoke and from exhaust gases, amongst other things. So we need further research into smokeless fuels, into catalytic converters. But that has nothing to do with carbon dioxide, which is a food for plants, which in turn produces oxygen for us to breathe. How many times do we have to repeat the first steps in Biology concerning photosynthesis?

O Princes, O Lords Divine and temporal, for whom do the bells toll? They toll for all mankind.

Anthony Bright-Paul

14th of December in the Year of Our Lord 2009.

What is Global Warming?

Anthony Bright-Paul

Tuesday, February 09, 2010.

Since so much alarm has been created over Global Warming, one might have expected that there would be a clear definition of what Global Warming is, that everyone is agreed upon. Alas, such is not the case and much confusion results.

Just for the fun of it, when you next come across an avid Global Warmer, ask how they define Global Warming. What precisely is supposed to be warming? Not the magma, surely! The surface of the planet? Is the desert warming at the same time and rate as the Arctic ice? Or is it the atmosphere? Or, the near surface atmosphere? The troposphere as a whole? Or the mid-troposphere? In order to find a satisfactory definition I have had recourse to Google. So the first definition from Allianz is 'the increase of the average temperature on Earth.' Tell me, do you

find that a satisfactory definition? Down below you will find a picture of the worldwide temperatures at 8 am today in the UK. I defy anyone of you to take a meaningful average of the world temperature at that time.

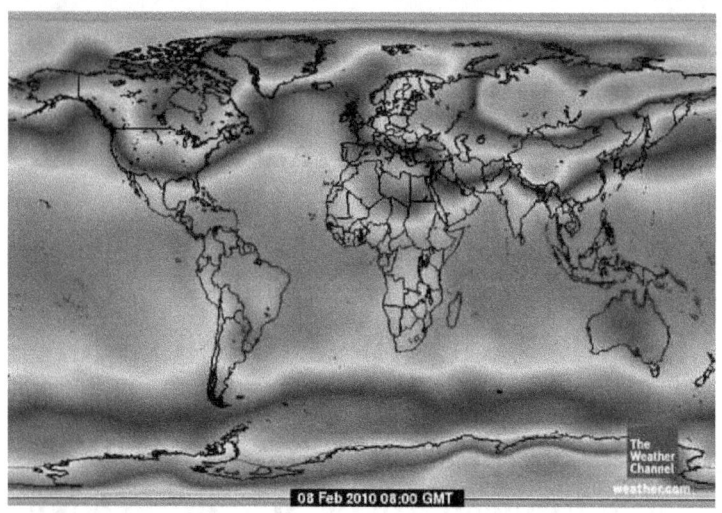

If for example we were to take the average of the North Pole at -34°C it balances nicely with Accra at +34°C. Are we any the wiser?

Of course the definition from Allianz has conveniently left out a time-frame. First of all an average is clearly a nonsense at any given time of day, and secondly it is a nonsense unless a given time frame is agreed. Should we take a week or a couple of months? Should we take a year, a hundred years, a thousand years?

If we take but a few days then we know that at the present moment Washington is covered in 2ft of snow, and is expecting another heavy downpour. The UK is enjoying the severest winter for years, as is most of Europe, starting with the conference on Climate Change in Copenhagen.

So while we in the Northern Hemisphere are perishing from the cold, while the so-called scientists and politicians wedded to the idea of Global Warming have been exposed for fixing the books, putting in false data and refusing to show their data- sets, there are those like Prince Charles and Ed Miliband, who bleat that the fundamental science has not changed!!

How very right they are!

That the sun's radiation is the prime driver of climate has not changed. The Earth's orbit has not changed, nor that wobble that is called the Milankovich Effect has not changed. Great Nature in its awesome unpredictability has not changed.

What needs to change is the phoney and militant hysteria of those who have latched on to a completely unproven theory that a trace gas in the atmosphere is causing this ill-defined Global Warming which is supposed to be causing an ill-defined Climate Change.

But let us return to the time frame, as the word 'increase' implies a time frame. Guess what? Nobody has agreed on one. In fact everyone it appears only uses the time-frame that is most convenient to their arguments. If we don't take the last quarter should we take the last thirty years? Or the years since the end of the Great War in 1945? Or from the beginning of the Industrial Revoltion? Or from the end of the Little Ice Age, about 1850? Or should we take the average of the last thousand years? Or ten thousand or a million?

You see what I mean. First we have to agree a time- frame, without which all is nonsense. And then we have to agree on the data, where it comes from and who is collecting it, and who collected it in the past.

Wow! You can see how much room for chicanery there is in all of this. The Avid Global Warmers will swear black and blue that the Planet is still warming, while we are freezing (I see snow flakes falling again just outside my window) and they may be right, depending on the time frame! Certainly the Globe has been cooling since 1998, but then 1998 was an exceptionally hot year – at least so we are lead to believe. The truth is that we cannot trust the figures from the Hadley Centre nor the GISS, following those revealing emails. In fact it is very difficult to get any reliable figures at all, since, surprisingly, the numbers of Weather Stations has declined sharply and the positions of many are in Urban Heat Islands.

Since beginning this article fresh data has come to light: -

[1] http://scienceandpublicpolicy.org/reprint/climategate_analysis.html

THE GLOBAL DATA CENTERS

Five organizations publish global temperature data. Two – Remote Sensing Systems (RSS) and the University of Alabama at Huntsville (UAH) – are satellite datasets. The three terrestrial institutions – NOAA's National Climatic Data Center (NCDC), NASA's Goddard Institute for Space Studies (GISS), and the University of East Anglia's Climatic Research Unit (CRU) – all depend on data supplied by ground stations via NOAA.

Around 1990, NOAA began weeding out more than three-quarters of the climate measuring stations around the world. They may have been working under the auspices of the World Meteorological Organization (WMO). It can be shown that they <u>systematically and purposefully, country by country, removed higher-latitude, higher-altitude and rural locations, all of which had a tendency to be cooler.</u>

Apparently the numbers have fallen from some 6,000 to around 1,500! One might have expected the opposite, but this is clear evidence of double-dealing on a truly massive scale.

If we look at the definitions of Global Warming that I have taken from Google it is easy to see that in almost every case a Definition has been followed by an Opinion. I need a definition, as to what is being warmed, the surface, the near-surface atmosphere, the mid-air satellite measurement and so on. I have not asked for a cause or an opinion, so I will attempt below to separate one from the other, by striking through what is irrelevant.

What is Global Warming?

Global Warming is defined as the increase of the average temperature on Earth. ~~As the Earth is getting hotter, disasters like hurricanes, droughts and floods are getting more frequent. Allianz~~ (Unsupported opinion.)

an increase in the average temperature of the earth's atmosphere (~~especially a sustained increase that causes climatic changes~~) (Non sequitur)
wordnetweb.princeton.edu/perl/webwn

An increase in the earth's atmospheric and oceanic temperatures ~~widely predicted to occur due to an increase in the greenhouse effect resulting especially from pollution~~ (Unscientific, illogical and plain wrong.)
www.mdbc.gov.au/subs/The_River/glossary.html
The progressive gradual rise of the earth's surface temperature ~~thought to be caused by the greenhouse effect and responsible for changes in global climate patterns~~. An increase in the near surface temperature of the Earth.
...(Agreed, that some people have been taken in by propaganda to this effect, but it is only an unproven hypothesis, nothing more.)
www.natsource.com/markets/index.asp

the changes in the surface air temperature, referred to as the global temperature~~, brought about by the enhanced greenhouse effect, which is induced by emissions of greenhouse gases into the air.~~ Simply unproven.
citizenship.yara.com/en/resources/glossary/index.html

An overall increase in world temperatures ~~which may be caused by additional heat being trapped~~

~~by greenhouse gases.~~ -Yes, maybe and maybe not!
forecast.weather.gov/glossary.php

~~The term given to the major consequence of the greenhouse effect. Scientists have long predicted and recently measured notable increases in the world's temperature.~~ ...Some scientists have created a totally false picture by the infamous hockey stick.
www.emissionstatement.com.au/Climate_Change_Glossary_of_Terms.html

A process that raises the air temperature in the lower atmosphere ~~due to heat trapped by greenhouse gases, such as carbon dioxide, methane, nitrous oxide, CFCs and ozone.~~ ... (Really? I thought the sun was responsible for heating the lower atmosphere.)
www.nada.org/green/getinvolved/glossary/

an increase in the average worldwide temperature ~~primarily caused by fossil fuel burning and an increase of carbon dioxide in the atmosphere. (~~ Completely unproven theory ,disputed by some 19,000 top climate scientists.)
www.woodrow.org/teachers/esi/1998/r/plankton/gloss.htm

Global warming is the increase in the average temperature of Earth's near-surface air and oceans *since the mid-20th century and its projected continuation.* (Wikipedia)

Here at last is a definition with a time-frame, since 1950. Geologically speaking that is a very short period of time! As if the history of the Earth had never happened!

In the examples of definitions I have left in the definition and I have struck through the unproven opinions. There is no proof whatsoever that the average temperature worldwide is increasing at the present time, since it is appears that the CRU and GISS have cooked the books in order to give that impression. However if we take the depths of the Little Ice Age to the present day then surely there has been an agreed warming of about 0.7°C. On the other hand if we started with the peak of the Mediaeval Warm Period, then by the same measure the Globe is cooling.

With huge snowfalls in North East America, with the extreme cold killing off the flocks and herds in Mongolia, with avalanches in India, with continuing cold in the UK, it is as if the Almighty Powers above are trying to send mankind (manusia) a message that the Globe is not warming but is presently cooling quite rapidly.

Then let us look at the opinions that I have struck through. '...primarily caused by fossil fuel burning and an increase of carbon dioxide in the atmosphere.' Well, that is complete nonsense for a start. At 385 parts per million the total amount of CO_2 circulating in the atmosphere is miniscule. While this miniscule amount may have increased slightly in the last fifty years, the Global Temperature, if we are to trust the figures, has fallen. There is no correlation between one and the other. Thank God in His Heaven Above for Carbon Dioxide and Photosynthesis, without which we would not be able to breathe Oxygen, the breath of Life.

Then we have '...the additional heat trapped by Greenhouse Gases.' But there is no Greenhouse, there is a continuum. There is a theory, but there is no proof. Greenhouse gases only delay the exit of the sun's heat into outer space.

Then I like this one: '...a sustained increase that causes climatic changes.' Really? Even a child can see that this is a *non sequitur*. Climate is always changing, and always has done. The causes of climate change are manifold. To attribute it to a trace gas is simply nonsense. But then again we need to define Climate!

If we go to Google again and look up the Köppen-Geiger classification of Climate, we can count some 29 different zones. I suggest that everybody does just that, go to Google and choose for yourself the best images of the 29 zones, with nine major categories.

Köppen climate classification From Wikipedia, the free encyclopedia

Jump to: navigation, search

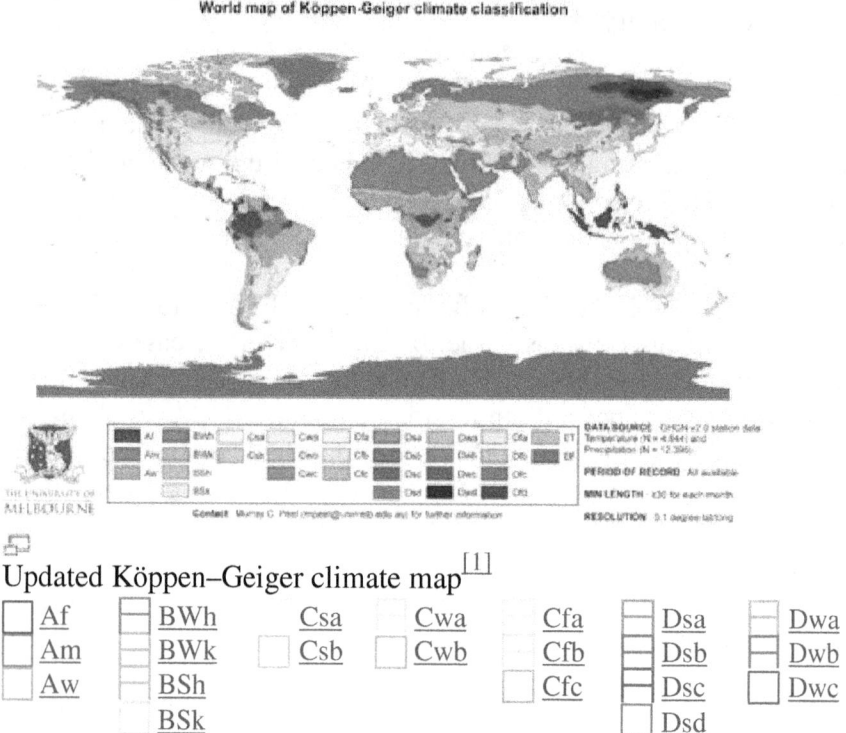

Updated Köppen–Geiger climate map[1]

Af	BWh	Csa	Cwa	Cfa	Dsa	Dwa
Am	BWk	Csb	Cwb	Cfb	Dsb	Dwb
Aw	BSh			Cfc	Dsc	Dwc
	BSk				Dsd	

Well, I have saved you the trouble. Just click on this map and you will be taken to the site. It exposes those who glibly talk of Climate Change, based on a few pictures and backed by the hysterical outpourings of charlatans and the weak-minded. In fact and in our own direct experience, there is not one climate, but at least nine main zones: Equatorial, Desert, Mediterranean, Sub-Tropical, Oceanic, Continental, Temperate, Sub-Arctic and Ice-Cap, with some twenty-nine sub-divisions.

These main divisions are immediately recognisable, helped by the coloured map. Clicking on the map itself will provide more information. The hot desert regions in the centre immediately stand out, while the dark grey shows the Ice -cap and the Sub- arctic regions. The UK shares a temperate climate with New Zealand.

Climate is immensely complex. To imagine that climate change is brought about by a trace gas that is 0.04% of the atmosphere is sheer folly if not downright stupidity. It is as if a person was unaware of the news that is broadcast every day.

Last night Sky News reported that the severity of the winter in Mongolia has caused the death of some 20 million livestock. Huge storms in North America have not only blanketed Washington in snow, but more importantly perhaps brought down power lines so that individuals who could work from home are deprived of their computers, let alone other home comforts. An avalanche in India killed 150 soldiers.

Everyone knows about the huge earthquake that has brought disaster to Haiti, but in addition there have been a large number of earthquakes in the last seven days. Just look up Recent Earthquakes. There have been over 60 earthquakes in the last week, with many of them over 5.0 on the Richter scale. In case that does not shake you please follow that by searching recent volcanic activity. And then again look to the Atlantic Jetstream that flows at 35,000ft., above our normal winds and weather systems.

The Global Warmers are inclined to jibe that the Skeptics are Deniers – that they deny that Global Warming and Climate Change is man-made. What a ridiculous accusation! On purely linguistic or semantic grounds the accusation is absurd without a clear and agreed definition of Global Warming and of the word Climate.

But when one thinks of the movements of the plate tectonics, the weekly number of earthquakes and volcanic eruptions, the unpredictable jetstreams and wind patterns and barometric pressure, the ocean currents and last but not least the solar activity, the very idea that all or any of this can be attributed to man, is just plain nonsense. Furthermore there are hot springs and geysers, and hot vents in the oceans, all of which have bearing on climatic matters.

The idea that all of this is man-made is so ludicrous as to be an insult to intelligence. Yet there are those who still persist that the ever-changing climate is anthropogenic. This belief, that has been pedalled assiduously by Al Gore and his henchmen, is a dangerous venom poisoning the body politic throughout the entire world. Unfortunately huge resources are marshalled to fight a miasma.

Could anyone of the Global Warmers create an earthquake? Can Prince Charles or Ed Milliband tear the tectonic plates apart? Can our Prime Minister or the President of the United States re-order the direction of the Atlantic Jetstream? Can Pachauri of the IPCC command the Gulf Stream or La Nina? Can any single one of these create a single cloud or produce a drop of rain? Yet they imagine that by limiting the amount of man-made Carbon Dioxide they will abort the Weather systems that go to make up Climate. They have simply not done their sums. If they realised just how vast is the Tropsophere, the lowest level of the atmosphere, they would surely think again. The volume of the Troposphere alone is 7.67 billion cubic kilometers!!! Surely it is they who are the Deniers, who deny the truly awesome power of Great Nature.

When we answer these questions it becomes clear beyond peradventure that man has never and can never produce Global Warming or Climate Change. This fiction has been brought about by those with no innate reverence for the Great Powers that have always ruled the Earth, the Moon, the Stars, the constellations and the Universes. Who are the real believers? And who are the real Deniers?

Anthony Bright-Paul
Tuesday, February 09, 2010

Postscript: While bitter cold and snow continues in the UK and is forecast for the coming week, last night we were shown horrifying pictures of roaring floods and mudslides in the island of Madeira, which has claimed the lives of more than 30 people. A swarm of earthquakes has hit Southern California.

I am happy to solemnly declare and to deny that any of these events and many more across the Globe has anything whatsoever to do with man.

Global Warming and that 2nd Law

Anthony Bright-Paul
March 30th 2010

I was resting and suddenly the 2nd law of thermodynamics came into my mind. As is my wont I had to go to Google to clarify this law – to make it comprehensible to a non-scientific mind. Really it is quite simple.

If I set light to my bonfire it will rage with a fierce heat and then it will subside and eventually even the ashes will go cold. But why is that? Why does the fire simply not get hotter and hotter? Well, we all know the answer without even knowing the Physics. We all know that heat travels from hot to cold, but cold never travels spontaneously from cold to hot. It is as simple as that.

If I plug in my electric iron and throw the switch the resistance to the current will make the iron hot. If I switch the current off, does the iron get hotter and hotter? Of course, not! Any simpleton knows that. But why not? Why does it not get hotter and hotter and reach a tipping point? And get so hot that everything melts? But it does not happen - the iron gradually cools.

If I fill my electric kettle full of water and switch it on, it will take less then 4 minutes to come to the boil, when it will switch itself off. The kettle and the water will gradually cool down, the water retaining its heat for a long time. But it will cool, and it will cool by the 2nd law of thermodynamics.

The *Flying Scotsman* used to run between London and Edinburgh non-stop within 8 hours. How did it do this? It had a tender containing 9 tons of coal, and a special corridor to allow for a relief driver and fireman. You may imagine that poor fireman shovelling coal non-stop perhaps for 4 hours, before getting relief. Without the coal the fire would have died down and insufficient steam would have been produced. The heat was only maintained by the continuous addition of fuel.

We humans are also combustion units. We need sufficient oxygen in order to process the fuel we continually imbibe and eat. Without food and oxygen the fire would soon go out and we would be cold and stiff.

Once we understand this principle it is really quite frightening. Even a nuclear power station only has a life of 25 years. What about the sun? An awful thought came to me and I had recourse to Google once again. If combustion depends upon sufficient fuel, then will the sun itself burn out over the course of time?

The answer is Yes! In about 5 to 7 billion years the sun will have run out of its fuel and it will burn out. The exact scenario will doubtless be dramatic. After all the sun is a star and it is subject to the same laws of Physics.

Where does that leave us? Where does that leave us in respect of Global Warming? Well, the answer is simple – there is no such thing as Global Warming. There is only Global Cooling by the inexorable laws of Physics.

In fact we all know this. We all know that we on this Planet Earth are kept alive by that huge fireball in the sky. We do not have to be a scientist to feel the warmth of the sun as it rises in the morning. This huge boiler in the sky gradually heats us up as the Earth spins round. As night falls the Earth cools on that side farthest from the sun. The amount of heat that is received depends upon the angle of the Earth to the Sun. So we know that the tropics enjoy most of the heat, the temperate climes a fair amount in the summer months, and the Poles remain pretty cold year in and year out.

What else do we know? We know that it is by a divine providence that we are at just about the right distance from the sun and with sufficient mass. We know also that we have an atmosphere; an envelope of gases, and this enables us and all life, animal and vegetable, to exist on the Planet. We know that 99% of the atmosphere is composed of Nitrogen and Oxygen, but a crucial 1% is composed of what are called the Greenhouse Gases – mostly Water Vapour 95% and tiny amounts of Carbon Dioxide, Methane and other minor trace gases.

Why are they crucial? Without these gases we would all fry by day and freeze by night. In fact that is well illustrated by those very dry regions like Morocco. The absence of water vapour means that they suffer extremes of heat by day, and extreme cold by night. One does not have to be a scientist to experience this. Neither does one have to be a scientist to experience the hot nights of Jakarta on account of the extreme humidity. Water vapour delays the exit of the sun's heat from the Planet.

Let me ask you a simple question, or rather let me ask myself a simple question. If by some freak of nature, if by chance we were knocked off course by a comet and the sun did not rise, would the Greenhouse gases by themselves make us hotter? Would they keep the world warm?

You know the answer, and I know the answer. The gases in the atmosphere do not heat anything. The atmosphere only warms up from the heat of the sun, as do the waters of the oceans. Without the mighty sun we would get gradually colder and colder.

There is no such thing as Global Warming!!! There is only Global Cooling.

The Clausius Statement:

<u>Heat</u> generally cannot flow spontaneously from a material at lower temperature to a material at higher temperature.

Hi Anthony:

I like this article because it achieves its goal. The only small quibble I have is with the concept of "fluke". I think coincidence is a better word.

Tim Ball

Weasel Words

Anthony Bright-Paul
November 16th 2010

From all sides we are assailed by weasel words and by weasel phrases. These are words and phrases designed to have an emotional impact on people of low intelligence.

The word 'intelligence' derives from the Latin word, of which the principal parts are: Intelligo, intelligere, intellexi, intellectum. It is not difficult to see that the words, intelligent, intelligence, intellect and intellectual derive from this Latin root, which means understanding and discernment.

Intelligent people use language carefully, weighing up the meanings of the words and phrases they use, and being fully aware when they are slipshod and use clichés sometimes for the sake of speed and convenience. Using words carefully so that the words relate to specific meanings takes a lot of time and trouble and self-discipline.

On the other hand despots (which unfortunately includes almost the entire political class) have scant regard for meaning, but only for the emotional effect of the words and phrases that they use. And why is that? For a very simple reason – despots have long understood that men are not moved by reason, but they are moved by emotion. In fact the very word emotion has a Latin root. That is why that excellent book by the former Chancellor of the Exchequer, Nigel Lawson, "An Appeal to Reason" is doomed by its own title. (Correction: 'doomed' is the wrong word, since I understand the book has been a best seller!) Who cares about reason, when emotive words and phrases can stir men and women to all sorts of unreasonable actions, unreasonable opinions and unreasonable emotions?

Powerful people, those who seek to have power over other people, do have a very low form of intelligence, which enables them to understand what words and phrases, what words used in certain contexts, can move men to do what they want. That is not to say that their supposed motives, and what they believe are their motives, are not for the good of mankind. We can simply think of the word 'Change' which used within a certain context was powerful enough to get a President elected to the White House. Now some millions of American citizens are wondering what did he really mean by this supercharged word. For nothing very much has changed for the poor, the blacks and the unemployed. No one can doubt his sincerity in using it, but one can doubt whether he himself really understood or wanted to understand the significance of what he was saying.

Our own Prime Minister lectures and hectors the Chinese on *humanrights* and also on *climatechange*. I have deliberately joined these two phrases up, as they are typical examples of weasel phrases, which have contracted into weasel words. I won't dwell on *humanrights* too much, for the truth is that we humans have not so much rights as obligations – as J.G.Bennett, author of "The Dramatic Universe" and "The Crisis in Human Affairs" used to say, 'We have to pay the debt of our arising.' I will not argue that one at present, for the concept that we owe something simply because we are born into this world, is at present a highly unfashionable concept. Nevertheless it is the truth.

But when we come to *climatechange,* which has morphed from a phrase into a single word concept, then we can more easily ask just exactly what meaning this concept has. It is very easy to see how emotive this phrase has become, and the more emotive it has become the more it has been stripped of any meaning. If you don't believe me then watch those who splutter with rage when this is put to them. Here is a Conservative Prime Minister of apparently good education, and certainly one cannot doubt his goodwill, yet he uses a phrase on the world stage that is clearly devoid of any real meaning.

How dare I say that *climatechange* is a phrase devoid of any real meaning? Easily! I defy any one of you to come up with a comprehensible definition, which is satisfactory to any group of men of intelligence and discernment. And how can I, an unknown old man, dare to propose such a thing? Because I am confident that if we were to put this to one hundred members of Parliament of all or any Party, that we would be given one hundred differing definitions.

First of all what is the meaning of the word 'climate'? Wherein lies the difference between weather and climate? Just how much weather makes a climate? How many climates are there on this Earth? How many microclimates are there on this Earth? Is there such a thing as a Global Climate? Is there such a thing as Global warming, another weasel phrase that is bandied about? Think about it. The Globe is a big place. Is the Globe warming? Come on now! Be real! Be reasonable, if only for a couple of minutes! I am willing to bet that if you asked 20 MPs and twenty members of the US Senate, and 20 members of the Australian Parliament, and the Canadian and the German and the French – I am willing to bet that you would come up with 20 differing understandings in each and every case.

Unfortunately these weasel words are legion, and they are used with abandon and with no respect for logic. So we have sentences like 'climate is changing **because of climate change'**. Surprise! surprise! If you think I have made that up here is a quote from the Sci-Tech section of the BBC. The marine ecosystems for the jellyfish have been destabilized **because of Climate Change.** In fact almost every day somewhere on the BBC we hear that something or other has occurred **because of climatechange!** One might just as well say, **Because things have changed they have changed.**

Another weasel word that has crept in is 'sustainable', when intermittent would almost certainly be more correct. 'Renewable' is in the same category.

Does it matter that there are such weasel words? That depends on your point of view. If you believe it is right to incite people, especially young people, to violence by the use of such words and ideas; if you believe it is right to dupe people, especially gullible people of all ages, film stars, old ladies and the like, into entering upon an idiotic crusade to **SavethePlanet** from emissions of CO_2; if you believe it is right to perpetrate such a howler as your **CarbonFootprint,** then for you it does not matter. In fact these weasel words serve the purpose of awakening in people the lower forces, which deprive them of true initiative.

So emotive have these weasel words become that scientists who simply question anthropogenic global warming receive death threats, are deprived of their jobs, have their businesses burnt down, have their grants stopped and so on and so forth. They are denied access to scientific publications, where the whole peer review process has been corrupted.

If this is correct, how then does this unwholesome juggernaut of distorted science and half-truths and downright lies keep rolling?

Watch this space!

Beware the global warming fascists:

By JOHNNY BALL

UPDATED: 09:03, 22 February 2011

At London's Royal Court Theatre last week, a new play opened to rave reviews. It's about an environmental scientist who — horror of horrors — doesn't believe in global warming. The play is called The Heretic and, though I haven't seen it yet, I could already sink to my knees in gratitude - because in my own quiet and reasonable way, I am that global warming heretic.

In the past decade or so I've been mocked, vilified, besmirched — I've even been booed off a theatre stage — simply for expressing the view that the case for global warming and climate change, and in particular the emphasis on the damage caused by carbon dioxide, the so-called greenhouse gas that is going to do for us all, has been massively over-stated.

Blinded, maybe even brainwashed by the climate-change zealots, we are spending so much money on reducing carbon emissions that there is a danger of us bankrupting ourselves — and future generations — to solve a problem that in the opinions of a growing number of scientists and opinion-formers has been wildly exaggerated.
To put it another way, those who have been worshipping so ardently at the altar of reduced carbon emissions — and how quickly they adopted the messianic zeal and intolerance of a religion — may find that they have been deifying not just a false god but a ruinously expensive one, too.

As someone who has dedicated his life to popularising science and mathematics for young people, I find it hard — hurtful even — to be cast in the role of villain. I'm also aware that many of the people who have been kind enough to enjoy my TV programmes over the years are surprised to hear me — nice, cheery, Johnny Ball — expressing such strong and arguably provocative views. So let me explain how I came to them.

A quarter of a century ago, when I was churning out television series almost twice a year, I hit upon a successful way of working. For every series of six science shows, I would come up with seven ideas — six plus a spare — and every one of those shows, including the spares, got made. Except for one: about renewable energy, a subject then very much in its infancy.

I instinctively warmed to the almost Heath Robinson-like engineering behind those early attempts to harvest the energy of the wind and the waves, the tide and the sun. But there was a big problem: Hard as I tried, I couldn't make the sums add up. These devices either didn't produce anything like enough energy, or the energy they produced was too expensive to be economically viable.

None came close to reproducing the power of the physical process that has driven our civilisation since the Industrial Revolution — the heating of water. Be it by charcoal, coal, gas or nuclear energy, it's warmed until the temperature approaches 100C, and it suddenly expands to become power-generating, engine-driving steam. Until renewable energy could do something like that, I couldn't see how it had a serious future. I checked the sums — they were right — and walked away. The show never got made.

Which would be completely unimportant but for one thing. In the 25 years since, the 'science' of global warming and climate change, with its carefully selected graphs of rising temperatures, sea-level and CO_2 levels (by the way, other carefully selected graphs can show the exact opposite) has all but conquered the world, and no opposing view is to be tolerated.

The cult of reducing carbon emissions shapes everything we do, at local, national and global levels. The very future of the planet, we are told, hangs on our dispensing with fossil fuels and adopting renewable energy sources as quickly as possible. But here's the problem — 25 years may have passed since I tried to make that TV programme and the technology has improved a little, but the sums still don't add up.

I'm quite sure renewable sources have a minor contribution to make to our energy needs, but they still don't produce anything like enough energy at anything resembling the right price to offer a viable future. **If it costs 2.3p to produce one unit of electricity using gas, it costs 2.5p to produce the same electricity using nuclear energy and perhaps 2.9p using coal. Using wind power, the cost is an astonishing 9.8p.**

In the face of such figures, most reasonable people interested in cleaner, sustainable energy would surely go off and build carbon-free, nuclear power stations or gas-fuelled ones. But advocates of global warming with their dire warnings about the evils of CO_2 emissions have got too firm a hold, their thinking become too widely accepted, for anything that sensible to be an option. Instead, they're changing the sums, and manipulating the maths.

The result is a growing burden of green taxes, renewable energy subsidies and unseen charges that will cost us — and particularly our children — billions and billions of pounds. Already, these additional costs are adding 50 per cent to all our energy bills, and 50 per cent to air-fares. At a time of severe economic hardship, when thousands of jobs are being lost and households struggle to make ends meet, this is a potentially ruinous burden.

But it's also — in my opinion and that of a growing number of others — an unnecessary burden. I'm all for the careful use of the Earth's resources and I applaud the many breakthroughs that have been made both in the recycling of these scarce resources and in the battle against environmental pollution. I want a clean, green planet. But this obsession with controlling carbon dioxide levels in the atmosphere is now as dangerous as it is ridiculous. Along with water and oxygen, carbon dioxide is one of the three basic requirements for sustainable life. And yet this natural gas — only 4 per cent of which is produced by man — has been branded as the greatest threat to the future of this planet. Well, forgive me, but I think that's nonsense.

My own view, for what it's worth, is that the water content of air has far more impact on temperature than carbon dioxide levels do. It's a common-sense belief based on simple observation — we all know the impact that a layer of cloud has on temperature — and basic chemistry that tells us that when we burn any sort of fossil fuel two molecules of water are produced for every one molecule of carbon dioxide.

But any increase in air temperature produced by raised water vapour levels will be minor and largely self-regulating. So climate change is absolutely normal and in my lifetime has never yet resulted in any signs that should cause alarm.

Already I can hear those wholeheartedly committed to the global warming cause queuing up to cast the first stone at such blatant heresy. But should we trust them? Not if they include the likes of Rajendra Pachauri, head of the UN-backed Intergovernmental Panel on Climate Change, who last year was forced to admit the panel had exaggerated the rate at which Himalayan glaciers were shrinking. Or, Dr Phil Jones, the Cambridge climatologist at the centre of a scandal after he was accused of manipulating and suppressing climate change data.

Or the scientists who realised that a 0.7C rise in global temperatures over the past 100 years was hardly the stuff of environmental Armageddon, so they looked for a particularly chilly year against which to compare today's figures, and found 1961. Suddenly, global temperatures had gone up at 0.7 degrees in just 50 years: now, that was more like it. It's statistics like these that give rise to the sort of absurd pronouncement we saw at the Copenhagen climate change conference late in 2009, where it was grandly announced that at a cost of $100 billion a year we might just be able to limit the increase in global temperatures to 1.5C by the end of the 21st century.

I regard that as an almost unimaginable amount of money effectively being poured down the drain, taking the futures of millions of young people with it. And that makes even a mild-mannered chap like me very angry.

I've had the good fortune to live in a period when things have always got progressively better, where each ten years is always a bit better than the previous ten years, where children can grow up pretty safe in the knowledge that they will enjoy better lives than their parents. Until now, when we're suddenly told it all has to stop.

I don't make television programmes any more, but I do still visit 80-100 schools a year and I know what children are taught about climate change, and what the result is. They accept it absolutely and will solemnly tell you that they always turn off lights, close doors and, at school, have installed solar panels on the roof. They tell me how worried they are about global warming, rising sea-levels and, having seen alarmist films such as Al Gore's An Inconvenient Truth, the imminent prospect of all human life being wiped out.

And this breaks my heart. I want children to be excited about the future, not cowed by it. I want them to grow up in a world which is going to be better than the one their parents knew, not significantly worse. I want them to grow up excited by technology and new inventions, not worrying about where the electricity is going to come from to power them.

Efficient: Power stations are as working at twice the level they were 18 years ago and their safety record is second to none.

And that exciting future could still be theirs. There is plenty of power out there if we'll only let them have it. Gas-powered power stations are now twice as efficient as they were 18 years ago, while the safety record of nuclear power in Western Europe is second to none. We ought to be investing in these, not ridiculous and highly inefficient wind farms that are only being built because of the huge government subsidies and guaranteed profits that are being offered.

People have a right to know the truth, but it's so difficult to break the strangle-hold the global warming gang have on the debate. David Bellamy can't get on television and I can't even get a ten-minute meeting with the controller of Radio 4. Maybe that doesn't matter, maybe our time has passed. But what does matter is that the view held by the so-called climate change sceptics must be heard too.

We're not going to hell in a handcart because of some tiny increase in atmospheric CO_2 levels, but hell is certainly where we'll end up if we insist on spending hundreds of billions of unaffordable pounds trying to correct a problem that is grossly over-stated. It's time to get back to the real science and the real sums. It's time to get back to the future that, if we adopt the right policies, will be brighter than we can yet imagine.

My Lack of Physics

Questions: and answers by Hans Schreuder

Anthony Bright-Paul
12.03.2011

There is something that still puzzles me. If the sun is a great ball of fire it is radiating out heat. If that is the case then the nearer you are to the sun you should be hotter.

Correct. But that's only because you or the vehicle in which you travel has mass to it.

Therefore outer space should be really hot. But I understand that it is really cold. Why?

Two interlinked errors here....
1. Outer space is a vacuum and thus has no mass within it to have a temperature. Any vacuum is devoid of matter and thus devoid of a temperature ... temperature is a measure of the flow of energy.
2. As per 1. above, outer space is not freezing cold either....! It has no temperature! The confusion comes in with the concept of a Cosmic Background Radiation, which is in the order of 3K (three Kelvin, equivalent to minus 270°C). That CBR is "measured" by looking through trillions and trillions of miles of the vacuum of space and thus the odd molecule of matter that really does have that 3K "temperature" ends up being registered on super sensitive IR "thermometers". Various words are in "" because they do not fit the usual meaning of those words.
To add to this confusion, any matter in that vacuum of outer space and not in receipt of any stellar (solar) radiation either is or will get colder and colder until it reaches that same 3K. Why? Because all matter radiates when it has a temperature above zero K, the absolute zero point of all matter, and the point where no further radiation is possible. So, in the absence of stellar (solar) radiative input of energy

(in our case sunlight) any matter in outer space will continue to lose its energy by radiating until it gets to the CBR level of 3K.

The sun's rays pass through the atmosphere. Correct?

Not all of it. Approximately 25% of all sunlight that reaches our earth's outer atmosphere (a poorly defined region between 30 and 100km off the surface) is either reflected straight back into outer space off particulates (dust) in the atmosphere or the tops of clouds or scattered by infrared reactive gases such as water vapour and carbon dioxide and to a small extent the oxygen and ozone molecules as well. See this graphic to get some idea of the losses and the "absorption bands":

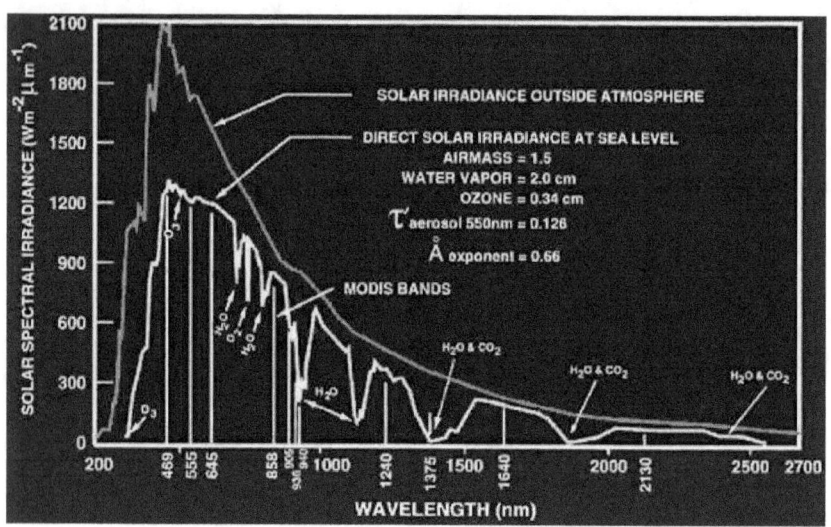

The top of our atmosphere, the Troposphere is extremely cold. The tops of mountains are cold. Yet the surface temperatures are hot, which are farthest from the sun. It seems illogical.

That has to do with the reactivity to radiation of the substances that are hit by that radiation. As per the previous point, some 75% of solar energy goes straight through the atmosphere and thus has no effect upon the air temperature. When that 75% of solar energy hits the solid earth or the water in the oceans, rivers and lakes, those substances are heated to varying degrees, depending on the specific properties of the material receiving that solar radiation. You will know from direct experience that a wooden bench feels less cold than a metal bench in winter yet warmer in summer. That's because wood has a lower specific heat. So, when it is cold, wood will feel less cold than a metal object yet warmer when the sun is out. This has also to do with the rate of heat dissipation, so not an easy concept to explain in a few words.

The atmosphere has no heat of its own. Correct? The sun's rays strike the ground. The immediate atmosphere becomes warm to a certain level. However if one sits in the sun one feels the heat from the sun. Is that an illusion? If a cloud passes it feels cooler.

Another tricky one to explain in a few words...! The atmosphere does have a temperature of its own but it has a low thermal capacity - example of thermal capacity: put a sheet of thin aluminium foil in the oven and put it on top of a ceramic dish. Let the oven be at 100°C only, so as not to create too many blisters ... Now, after 15 minutes or so, open the oven door and grab the foil with your bare hands.... I bet you will not get one blister and barely feel the heat of the foil; do not do the same with the ceramic dish, for obvious reasons! Explanation: the ceramic dish has a hugely higher thermal capacity. In fact, put that 100°C dish on a heat-proof surface, using oven gloves, and check its temperature after, say 15 and 30 minutes. Still warm, right? Yet that foil that came out of the same oven at the exact same temperature lost all of its heat in seconds. That's what's called thermal capacity.

So, back to the atmosphere. Air has a low thermal capacity, which is however greatly affected by the level of water vapour. The higher the humidity, the less the effect of a cloud passing overhead. Why? Because water vapour has a huge thermal capacity. So huge in fact that even nuclear reactors still use steam- driven generators to create the actual electricity they deliver! Gas, oil and coal-fired power stations all use steam-driven generators! That's also why the "age-of-steam" was such a gigantic leap forward in our industrial revolution. It is still used in our 21st century electrical generators!

If I light a bonfire it is very hot close to – scorching. Five yards away it is tolerable. 15 yards away I can see the flames but experience no heat.

Correct, that's due to a double-whammy effect: 1. convective heat-dissipation; 2. adiabatic heat-loss.

1. Any gas molecule that is only slightly warmer than its surroundings will have a lower specific gravity and thus become relatively lighter than its surroundings and thus floats away upwards.

2. Any gas that expands loses heat in so doing, called the adiabatic effect. That's why the higher up in the atmosphere you go the cooler it becomes - even though that is only half the explanation, as the atmosphere has distinctly warmer zones higher up, but that's another story and a different set of effects ... sorry to make it so complex, it is complex!

The only heat that you end up feeling directly after only a short distance away from the fire is the direct infrared radiation, which is distinct from the direct contact with the heated air.

Check this graphic to see how all known atmospheres have a kink in their temperature vs altitude correlation:

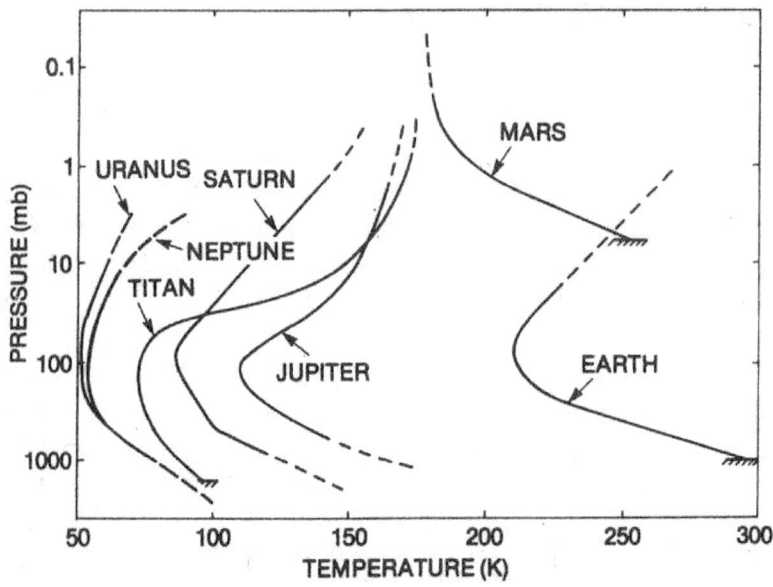

and to make it even more confusing, here is a close up of our own atmosphere:

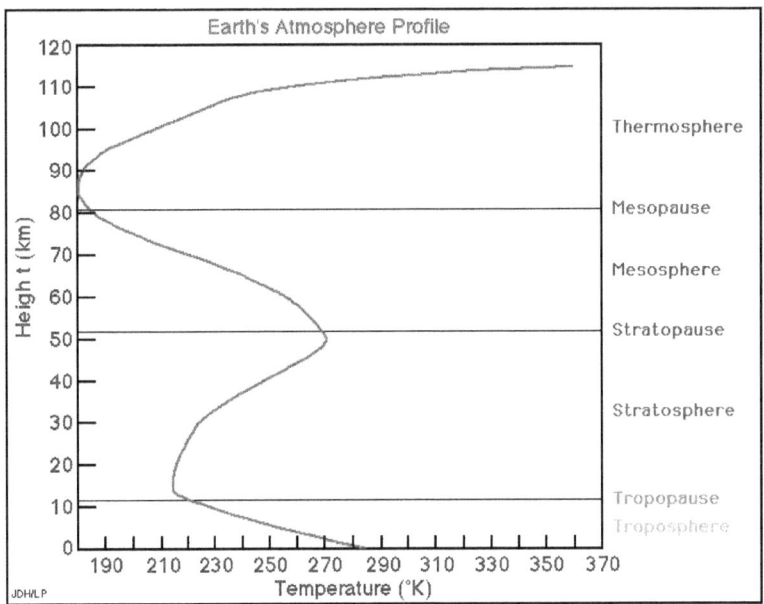

The sands by the Med get so hot to the feet that one has to wear sandals. Yet two hours after dusk these same sands are cold. Correct?

Yes, that is because the atmosphere will carry the heat away each and every second of the day, as per the above explanations. Also see my latest essay on that issue, attached. Sand has quite a high thermal capacity, that's why it takes a few hours for it to cool down. Hot rocks will cool down slightly slower, but cool down they will, even in the highest humidity zones. The only difference will be the rate of cooling, never the action of cooling as such.

Can you explain?

I've tried ... let me know if I have succeeded.

Global warming: 10 little facts

by Bob Carter March 14, 2011

Control the language, and you control the outcome of any debate

Ten dishonest slogans about global warming, and ten little facts.

Each of the following ten numbered statements reproduces verbatim, or almost verbatim, statements made recently by Australian government leaders, and repeated by their media and other supporters. The persons making these arguments might be termed (kindly) climate-concerned citizens or (less kindly, but accurately) as global warming alarmists.

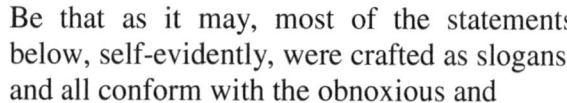
Bob Carter

Despairing of ever hearing sense from such people, some of whom have already attributed the cause of the devastating Japanese earthquake to global warming, a writer from the well regarded *American Thinker* has badged them as "idiot global warming fanatics".

Be that as it may, most of the statements below, self-evidently, were crafted as slogans, and all conform with the obnoxious and dishonest practice of political spin – in which, of course, the citizens of Australia have been awash for many years. The statements also depend heavily upon corrupt wordsmithing with propaganda intent, a technique that international Green lobbyists are both brilliant at and relentless in practising.

The ten statements below comprise the main arguments that are made in public in justification for the government's intended new tax on carbon dioxide. Individually and severally these arguments are without merit. That they are intellectually pathetic too is apparent from my brief commentary on each.

It is a blight on Australian society that an incumbent government, and the great majority of media reporters and commentators, continue to propagate these scientific and social inanities.

1. We must address carbon (sic) pollution (sic) by introducing a carbon (sic) tax.

> The argument is not about carbon or a carbon tax, but rather about carbon dioxide emissions and a carbon dioxide tax, to be levied on the fuel and energy sources that power the Australian economy.
>
> Carbon dioxide is a natural and vital trace gas in Earth's atmosphere, an environmental benefit without which our planetary ecosystems could not survive. Increasing carbon dioxide makes many plants grow faster and better, and helps to green the planet.

To call atmospheric carbon dioxide a pollutant is an abuse of language, logic and science.

2. We need to link much more closely with the climate emergency.

> There is no "climate emergency"; the term is a deliberate lie. Global average temperature at the end of the 20th century fell well within the bounds of natural climate variation, and was in no way unusually warm, or cold, in geological terms.

Earth's temperature is currently cooling slightly.

3. Putting a price on carbon (sic) will punish the big polluters (sic).

> A price on carbon dioxide will impose a deliberate financial penalty on all energy users, but especially energy-intensive industries. These imaginary "big polluters" are part of the bedrock of the Australian economy. Any cost impost on them will be passed straight down to consumers.

It is consumers of all products who will ultimately pay, not the industrialists or their shareholders.

---

4. Putting a price on carbon (sic) is the right thing to do; it's in our nation's interest.

> The greatest competitive advantage of the Australian economy is cheap energy generated by coal-fired power stations.

To levy an unnecessary tax on this energy source is economic vandalism that will destroy jobs and reduce living standards for all Australians.

---

5. Putting a price on carbon (sic) will result in lower carbon dioxide emissions.

> Economists know well that an increase in price of some essential things causes little reduction in usage. This is true for both energy (power) and petrol, two commodities that will be particularly hit by a tax on carbon dioxide emissions.

> Norway has levied a tax on carbon dioxide since the early 1990s, despite which a 15% INCREASE in emissions has occurred.

At any reasonable level ($20-50/t), a carbon dioxide tax will result in no reduction in emissions.

6. We must catch up with the rest of the world, who are already taxing carbon dioxide emissions.

> They are not. All hope of a global agreement on emissions reduction has collapsed with the failure of the Copenhagen and Cancun climate meetings. The world's largest emitters (USA and China) have made it crystal clear that they will not introduce carbon dioxide tax or emissions trading.
>
> The Chicago Climate Exchange has collapsed, chaos and deep corruption currently manifests the European exchange and some US states are withdrawing from anti-carbon dioxide schemes.

Playing "follow the leader" is not a good idea when the main leader (the EU) has a sclerotic economy characterised by lack of employment and the flight of manufacturers overseas.

---

7. Australia should show leadership, by setting an example that other countries will follow.

> Self-delusion doesn't come any stronger than this.

For Australia to introduce a carbon dioxide tax ahead of the large emitting nations is to render our whole economy to competitive and economic disadvantage for no gain whatsoever.

---

8. We must act, and the earlier we act on climate change the less painful it will be.

> The issue at hand is global warming, not the catch-all, deliberately ambiguous term climate change.

Trying to prevent hypothetical "dangerous" warming by taxing carbon dioxide emissions will be ineffectual, and is all pain for no gain.

---

9. The cost of action on carbon (sic) pollution (sic) is less than the cost of inaction. This statement is fraudulent. Implementing a carbon dioxide tax will carry large costs for workers and consumers, but bring no measurable cooling (or other change) for future climate.

For Australia, the total cost for a family of four of implanting a carbon dioxide tax will exceed $2,500/yr* – whereas even eliminating all of Australia's emissions might prevent planetary warming of 0.01 deg. C by <u>2100</u>**.**

---

10. There is no do-nothing option in tackling climate change.

> Indeed. However, it is also the case that there is no demonstrated problem of "dangerous" global warming. Instead, Australia continues to face many self-evident problems of natural climate change and hazardous natural climate events. A national climate policy is clearly needed to address these issues. The appropriate, cost-effective policy to deal with Victorian bushfires, Queensland floods, droughts, northern Australian cyclones and long-term cooling or warming trends is the same.
>
> It is to prepare carefully for, and efficaciously deal with and adapt to, all such events and trends whether natural or human-caused, *as and when they happen*. Spending billions of dollars on expensive and ineffectual carbon dioxide taxes serves only to reduce wealth and our capacity to address these only too real world problems. **Preparation for, and adaptation to, all climate hazard is the key to formulation of a sound national climate policy.**

Professor Bob Carter is a geologist, environmental scientist and Emeritus Fellow at the Institute of Public Affairs

Standard Atmosphere with Hans Schreuder on Radiation

Feet	Miles	Kilometres	Centigrade	Fahrenheit
32000	6.06	9.75	-49	-56.00
31000	5.87	9.45	-47	-53.00
30000	5.68	9.14	-45	-49.00
29000	5.49	8.83	-43	-45.00
28000	5.30	8.53	-41	-41.50
27000	5.11	8.22	-39	-37.80
26000	4.92	7.92	-37	-34.10
25000	4.73	7.61	-35	-30.40
24000	4.54	7.30	-33	-26.70
23000	4.35	7.00	-31	-23.00
22000	4.16	6.69	-29	-19.30
21000	3.97	6.39	-27	-15.60
20000	3.78	6.08	-25	-11.90
19000	3.59	5.77	-23	-8.20
18000	3.40	5.47	-21	-4.50
17000	3.21	5.16	-19	-0.80
16000	3.02	4.86	-17	2.90
15000	2.83	4.55	-15	6.60
14000	2.64	4.24	-13	10.30
13000	2.45	3.94	-11	14.00
12000	2.26	3.63	-9	17.70
11000	2.07	3.33	-7	21.40
10000	1.88	3.02	-5	25.10
9000	1.69	2.71	-3	28.80
8000	1.50	2.41	-1	32.50
7000	1.31	2.10	1	36.20
6000	1.12	1.80	3	39.90
5000	0.93	1.49	5	43.60
4000	0.74	1.18	7	47.30
3000	0.55	0.88	9	51.00
2000	0.36	0.57	11	54.70
1000	0.17	0.27	13	58.40
Ground			15	59.00

Pictures are often better than a thousand words. The Standard Atmosphere used by Airline Pilots illustrates very clearly that temperature declines with altitude, some 2° Celsius for every 1,000 feet.

I can't think in feet too easily, and even less easily in metres! But this chart makes clear that at 8,000 feet, which is one and half miles high, the temperature is already below zero C. Ascending further towards the Tropopause, where most airplanes now fly, one can see it gets progressively colder and colder. Here I repeat the chart so that it can be seen from top down.

Feet	Miles	Kilometres	Centigrade	Fahrenheit
32000	6.06	9.75	-49	-56.00
31000	5.87	9.45	-47	-53.00
30000	5.68	9.14	-45	-49.00
29000	5.49	8.83	-43	-45.00
28000	5.30	8.53	-41	-41.50
27000	5.11	8.22	-39	-37.80
26000	4.92	7.92	-37	-34.10
25000	4.73	7.61	-35	-30.40
24000	4.54	7.30	-33	-26.70
23000	4.35	7.00	-31	-23.00
22000	4.16	6.69	-29	-19.30
21000	3.97	6.39	-27	-15.60
20000	3.78	6.08	-25	-11.90
19000	3.59	5.77	-23	-8.20
18000	3.40	5.47	-21	-4.50
17000	3.21	5.16	-19	-0.80
16000	3.02	4.86	-17	2.90
15000	2.83	4.55	-15	6.60
14000	2.64	4.24	-13	10.30
13000	2.45	3.94	-11	14.00
12000	2.20	3.03	-9	17.70
11000	2.07	3.33	-7	21.40
10000	1.88	3.02	-5	25.10
9000	1.69	2.71	-3	28.80
8000	1.50	2.41	-1	32.50

7000	1.31	2.10	1	36.20
6000	1.12	1.80	3	39.90
5000	0.93	1.49	5	43.60
4000	0.74	1.18	7	47.30
3000	0.55	0.88	9	51.00
2000	0.36	0.57	11	54.70
1000	0.17	0.27	13	58.40
Ground			15	59.00

This is easily verified by anyone flying today, as most planes have a monitor which not only shows a map and the approximate location of the plane, but also the speed, the height and the outside temperature. Note that there is only a relatively small band where the temperatures are above zero, yet we have scientists who would persuade us that heat is not only trapped up there, but is reflected back to the surface of the Earth. That would confound a fundamental law of Physics, 'that heat by itself will only flow from hot to cold and never vice versa'.

At 20,000 feet it is already –25C and at 30,000 it is –45C. That is the temperature of all the gases present. There is no lurking hotspot, no conglomeration of Carbon Dioxide molecules sending spurious radiation back to earth.

The whole question of 'radiation' is an extremely complicated one, for which reason my mentor, Hans Schreuder, has warned me to tread warily, if at all. So I will quote him, in order not to get out of my depth, and for the good reason that I believe it will be educative for all.

Dear Tony,

One small but critical distinction between what I wrote and how you understand it ….

> If that is the case then the nearer you are to the sun you should be hotter.
>
> Correct. But that's only because you or the vehicle in which you travel has **mass to it.**

The reason I wrote that is because the *vacuum of space* that is between the sun and you has no matter in it; a vacuum is just that: it has nothing in it, there is no matter within it, thus there is nothing to make it hot (or cold for that matter).

Now that we've advanced somewhat, I need to add one further caveat to what I wrote in red:

But that's only because you or the vehicle in which you travel has **mass to it** *that is reactive to the radiation coming off the sun.*

The bulk of the gases that make up our atmosphere are not reactive to solar radiation and thus are mostly transparent to that solar radiation.

It's the *vacuum of space* that has no temperature, not the atmosphere, which has a whole range of temperatures.

And yes, here on earth, everything that is hot (=warmer than its surroundings) produces convection and radiation.

In space there is no convection, only radiation. The International Space Station has a major problem getting rid of its excess heat and needs massive external structures of external pipes with pumped liquid ammonia to radiate that excess heat into the vacuum of space.

The radiation coming off the bonfire does indeed not go far *in terms of sending the heat of the fire over a distance so you can feel it*! This is tricky to explain! As per Dave Haskells's explanation of the Inverse Square Law, the further away from the source you are, the lower the *level of intensity* (usually called the flux density) of the radiation. So, even though that self-same radiation will go on till infinity, eternally, the *intensity* of it reduces to such an extent that you can no longer "feel it".

Imagine that you are feeling bitterly cold and approach a bonfire. You will feel the heat from a certain distance; now imagine that same bonfire on a hot summer's night .... you need to be much closer to the fire in order to feel the heat. Why?

The reason you "feel" the radiation is because your skin is at a lower temperature than the energy density carried by the radiation.

Hope the above helps in understanding the rather complex nature of radiation. And I only know the half of it!

Kind regards, Hans

The other half next week!!!

A final gem from Hans, who corrects me painstakingly: -

It can move heat even through empty space. Misleading! There is no "heat" in radiation. If that radiation does not encounter the right stuff, it will not *do* anything, it will not *heat* anything. Radiation is *not* heat! Yet ... radiation can *cause* heat and heat *produces* radiation - try and work that one out!

Anthony Bright-Paul
September 20th 2011

Two halves make one whole!

Anthony Bright-Paul
26 September 2011

Some things are so profoundly simple that it takes time for them to sink in. 'Why is the sun able to heat the Earth?' asks Alan Siddons in the first chapter of 'Slaying the Sky Dragon'. Because the Earth is colder.

Have you taken that in? **'Why is the Sun able to heat the Earth?' Answer: Because the Earth is colder.**

It seems so utterly simple, so obvious, and yet there are a whole raft of learned Professors and Scientists and savants, and journalists and politicians who do not, or who appear not, to understand this most simple proposition. Or, to be more exact, what follows from it.

Just imagine for one moment that the Earth was as hot as the Sun. Would the irradiance from the Sun then double the temperature on the Earth? Clearly not! Because if that were so the irradiance from the Earth would then double the heat of the Sun, which in turn would double the heat of the Earth, and with such incessant doubling we would all be hotter than Hell fire.

No. If the temperature of the Earth equalled that of the Sun, **nothing whatsoever would happen.**

Yet this is precisely what the Alarmist Warmers have failed to understand. Even some Sceptics fail this simple maths exam. They take it as simple arithmetic, plus a bit of arcane Physics regarding radiation and photons. I will come back to that later.

Here is an experiment. Take two calibrated jugs and fill them half full with warm water. Then empty the second jug into the first. Will you now get double the heat? Better still if you fill the jugs with water at half the temperature of boiling point, 50° Celsius, and then pour the second into the first, will you then get boiling water at 100°

Celsius? Of course, you should. Two halves make one whole. Half a cake plus the other half makes a whole cake. The maths is irrefutable. Yet it does not work. Where is the error?

It does not matter if you understand the maths or not, even a simpleton knows that there has to be a difference. If my cup of tea has stood a while and gone tepid, it is no good adding more that is also tepid. I boil up another kettle and add a little boiling water. A judicious addition will make the tea drinkable.

So now, let us come back to climate. The Alarmists argue that as the Sun heats the Earth, then the infrared photons bounce off the earth and the oceans and shoot back to Outer Space, but this infrared hits the molecules of Carbon Dioxide on the way which then radiate every which way, including downwards back to the Earth's Surface. This then <u>adds</u> to the mean surface temperature of the Earth.

Hullo! Can this be right? This simple addition?

Sorry! This cannot be right. Only if these molecules were hotter than the Earth and Oceans, only then could this possibly be right. If these molecules were exactly the same temperature then there would be no difference, and ergo no difference would be made. No difference whatsoever!

Once we add the Adiabatic Lapse Rate into the equation, once we realise the simple truth that the gases of the atmosphere get colder with altitude, then we know with absolute certainty that the whole theory of Anthropogenic Global Warming is not just flawed, but totally, absolutely and completely wrong.

Anthony Bright-Paul
26 September 2011

Nice article, Anthony. The non-technical public needs more straightforward talk.

I got a little nervous at first with your sentence **"Will you now get double the heat?"** You will, in fact, have doubled the heat *energy* in the full jug compared to a half-full one, but the total system energy (both jugs) has of course not changed. Thermodynamics is full of linguistic mousetraps with which one can get pinched if we don't choose our words carefully.

Every pilot is aware of the adiabatic lapse rate which describes the way temperature drops as we increase in altitude. This closely parallels what one would expect from cooling by thermal conduction, not radiation - which seems to be completely ignored in the current argument about IR absorbers like methane, CO_2, water vapor, etc.

In fact, neither the skeptics nor alarmists version of radiation theory explains why the ever-decreasing temperature suddenly stops, then reverses when we leave the troposphere and enter the stratosphere with warmer layers higher up and cooler layers farther down. This is in contrast to the troposphere near the Earth's surface, which is cooler higher up and warmer farther down.

I'd also argue to the general public that CO_2 is a pretty lousy "greenhouse gas", at any rate. It's a simple linear molecule, with no permanent dipole moment, and can only be "excited" in a single "bend mode" and then only when struck precisely by a IR photon of about 14.77 micron wavelength. Couple that puny absorption capability with an atmospheric concentration of only about 400 ppmV and pardon me while I die laughing at your computer simulations.

Best wishes, and keep up the good work.

Jim Peden
Atmospheric Physicist

Dear Tony,

You may walk tall from now on my good friend. Your way with words is hitting the high notes. Well done and glad to know you! Kind regards, Hans Schreuder

From Alan Siddons:

I admit that my eyes got a bit misty when I read those first few words! That's exactly the kind of insight I wish to impart to readers. I never heard of Anthony Bright-Paul before Hans mentioned him a month ago. But Hans mentioned him again just recently — and here he is again. His Two Jugs example is very apt: adding one 'warm' to another warm just gives you one warm. I pointed out the same thing a few days ago on Hockey Schtick.

Lindzen argues that 240 watts from the surface matched by 240 from the sky will make the surface radiate 480 watts per each square meter...

> If you aim your attention on temperature, however, the impossibility of such a thing becomes apparent. For Lindzen's scenario has a 255 Kelvin sky facing a 255 Kelvin surface — yet it is known that two bodies at the same temperature aren't able to transfer heat to each other. Nor, of course, can one of those bodies raise the other's temperature. 303 Kelvin resulting on the surface simply cannot happen, then.

Some people 'get' that and others don't. I'm pleased to know that Bright-Paul does. Thanks for forwarding that, Tim.

Alan

From: Tim Ball: This gentleman contacted me a few years ago as he wanted to learn about climate change. He has since communicated with others including Richard Lindzen. I have not heard from him directly for at least a year but this morning he sent this email. I assume he is sending it to Lindzen and others but don't know. I have asked him. Tim

Simple Arithmetic

Anthony Bright-Paul
December 2nd 2011

One hundred times one makes one hundred. One hundred times two makes two hundred, - but what does one hundred times nought make? It makes nothing. One thousand nothings make nothing. One trillion trillion nothings also make nothing.

This piece of simple arithmetic is very essential for understanding the Universe, Climate and the Atmosphere. A vacuum is full of nothing; it is void, it is empty. A vacuum, empty, surrounds our Earth. Even the highest reaches of our atmosphere are almost empty, such as the Thermosphere.

When Captain Joe Kittinger ascended some 20 miles up in a helium balloon on August 16th 1960 he had to wear special clothing to insulate him in order to keep warm. At over 102,800 feet it was some minus 80 degrees Celsius. On the other hand he had to wear a special reflective suit, otherwise his blood would have boiled and he would have fried to death within minutes. How can one explain such a phenomena?

Up there he was much nearer the sun, he was nearer the source of radiation. But radiation times nothing makes nothing. Radiation from the Sun is a great multiplier. Radiation times substance generates heat, heat that is quantifiable – that is temperature. So that any substance or mass that ventures into the thermosphere will be multiplied by the sun's radiation, which works, so my friends assure me, by an inverse square law.

Some learned works describe the Thermosphere as being very hot, as Wikipedia, which is a paradox. On the one hand it is deathly cold, precisely because there is virtually no matter there at all, no heat content. On the other hand, because the molecules have a high kinetic energy there is technically a very high temperature.

But Captain Joe Kittinger was very much matter, very much mass. That is why he had to wear a special reflective suit; otherwise he would have been a dead duck within seconds.

When he fell from this great height there was nothing much to break his fall, there were very few molecules, and so he fell at great speed, surpassing the speed of sound. When the atmosphere grew denser he was able to open his parachute, but it still took some 15 minutes for him to parachute to the ground. Clearly a parachute would have had little effect without there being some drag.

Mountaineers when they ascend Mt Everest have to take oxygen with them to survive; the air is so thin that they have difficulty in breathing. And yet the summit is barely half way up the Troposphere. As the molecules there are so far apart it is mighty cold there too. We know that planes flying at 32,000 feet will show some minus 50C on their monitors.

Radiation has no temperature. Heat has no temperature – it is a concept. Electricity has no temperature. But what is heated has a temperature. There has to be substance or mass of some kind or another to have temperature, to be multiplied by radiation.

Clearly the air, the atmosphere, has very little mass. As it gets nearer the surface of the earth, the gases get denser. There is dust and water vapour, which can heat and likewise cool rapidly. At any given point on the surface of the earth the near surface temperatures are varying second by second. Gases are volatile, gases are reactive.

As the Earth is rotating on its axis, so the heat generated by the sun's multiplication, rolls round the Earth like a Mexican Wave. Except that the wave is humped at the tropics and is barely a ripple at the Poles.

Water has much more mass than air. This can clearly be demonstrated. Anyone can walk easily say 100 yards to a local shop. Try walking in the sea with the water at waist high. A good athletic schoolboy can run the 100 yards in under 12 seconds (my son-in-law did it in 11.2 seconds). Try swimming 100 yards. A good swimmer would take ten times as long as a walker, some 120 seconds. And that is an interesting figure, since water has about 1,000 times the thermal capacity as air.

So water is clearly less volatile than air. It heats up more slowly but it retains its heat far longer. Bringing a full kettle to the boil, and observing how long it takes for the water to get back to tap water temperature can easily illustrate this. The principle is the same for the oceans and the lakes.

The earth is clearly denser than both the air and water. An artillery shell can travel through the air some 20 miles. A torpedo using the most advanced technology can travel say some 2 miles, and in a straight line with difficulty. A submarine under water can barely achieve ten miles per hour. Compare that with how long it takes man to bore a tunnel through a hill, or a mountain.

The Mt Blanc tunnel is just over 7 miles long. With the help of explosives and special drilling equipment and the efforts of over 350 men it took over 30 months, that is two and half years to build. That shows just how dense the Earth can be.

So it is clear to any normal observant person that the earth is denser than water and air, it has more mass. For that reason it takes longer to heat and longer to cool. Everything above zero Kelvin radiates heat, but the inverse square law governs the distance. So a good bonfire will radiate some 4 feet, but 10 feet away little or no heat is experienced. On the other hand heat rises by convection and cools as it rises.

Similarly the waters of the oceans, which warm slowly, retain their heat longer and also warm the near surface atmosphere, the most famous example of that being the Gulf Stream. The climate of the west of England and Scotland is noticeably warmer than the East side, precisely because of this effect.

Everything is further complicated by thermal capacity. In hot conditions with bright sunlight the tar on a road surface might melt. The sands by the Mediterranean in mid-summer are often too hot to walk on comfortably in bare feet. The steering wheel of an automobile may become too hot to handle, as happened to me many times when I was working in the San Fernando Valley back in 1961. On the other hand, while driving, the direct sun may make the driver really hot, but if he puts his hand to the glass of the side window it will be surprisingly cool.

So as the surface of the Earth warms, that includes the oceans, the Earth warms the atmosphere, by conduction and by radiation. Furthermore the crust of the Earth is but a thin crust over a raging furnace underneath. This is borne out by all the volcanoes, geysers and vents in the sea floor. It is said that the Deccan Traps in India were formed some 60 million years ago, with a series of eruptions that took place over 30,000 years. We dare not disregard the enormous internal heat of our Planet.

Again, some 80% of the world's volcanoes are submarine, under the sea. All this heat is rising up, dispersing itself into the atmosphere and cooling as it goes. Heat by itself is always flowing from hot to cold and never vice versa. This is a law of Nature - the 2nd Law of Thermodynamics.

The only way that the Earth gains heat is from the radiation of the Sun. And it only gains heat because it has mass. Simple arithmetic tells us that a trillion trillion nothings make nothing. If the Earth were void then no heat would be generated.

As it is, the Earth gradually loses its internal heat as it cools steadily. If the Earth does get warmer, as the Global Warmers would have it, it can only be because it has mass and through the radiation of the Sun, nothing else.

No gas, neither Water Vapour nor Carbon Dioxide, is capable in any way whatsoever of warming the Earth - in fact precisely the opposite. It can easily be demonstrated that the so-called greenhouse gases cool the Earth. Their temperatures are governed by the radiation from the sun, and the warmth that rises by convection and conduction from rocks, sand and water. The Earth is radiating heat into space, and were it not that the radiation from the sun is greater than the radiation from the Earth we would all soon perish from the cold.

As it is the Earth is a Being with an allotted life span. As a man gets older he runs out of energy, and as he runs out of energy he gets colder and colder till he dies. Both the Sun and the Earth have massively more energy than man and therefore their life spans will be comparatively greater. But everything that is hot is cooling and that includes man and animals, the Sun the Earth and the Stars.

Those who imagine that curtailing emissions from the burning of fossil fuels will somehow retard the heating up of this Earth are simply bonkers, living in cloud-cuckoo land. Those who imagine that man can control changes of climate are equally deluded. The Planet Earth is at a precise distance from the Sun that allows life on Earth. The Earth could wobble, or have a collision, or even turn over on its side for all we know. But to imagine that man could control climate, could control Global Warming, by limiting emissions of so-called greenhouse gases, is risible.

Heat is always leaking away. There is no Global Warming, only Global Cooling.

Anthony Bright-Paul
December 2nd 2011

With Acknowledgments to a series of great Skeptic scientists, who have guided and corrected me over the past 6 or 7 years. They know well who they are. Any mistakes in the above essay are mine and mine alone.

The Laws of Physics

Anthony Bright-Paul

March 28th 2011.

Do you suppose that before Newton saw that famous apple drop to the ground, do you suppose before he formulated the law of Gravity, do you suppose that people did not understand that law long before he defined it? Of course they did.

The Norman archers at the Battle of Hastings certainly did, as they shot their arrows high into the sky to fall with deadly effect right into the eye of poor King Harold. They knew with absolute certainty that those arrows would not fly forever towards the Heavens but would certainly fall back to Earth. They knew about gravity long before Newton. And indeed everyone who has fallen out of a tree knows the law. Every golfer who can drive a ball 250 yards knows the law – however hard he propels the ball upwards and forwards the ball inevitably falls back to the earth.

In fact this is the wonderful thing about most of the Laws of Physics. Although they may be formulated in the classroom, sometimes with fancy names, immediately that we understand them we find that they relate to the facts of our experience. We may not understand the mathematics, we may not understand the how and why, but we do understand quite clearly that they are that they are facts, that they are true.

The 1st, 2nd and the Zeroth Laws of Thermodynamics are a case in point. Do you imagine that an ancient smith in his smithy with his bellows heating up iron to put a ring round a wheel, or a shoe on a horse, do you imagine for one moment that he did not understand about conductivity? Of course he did – his life depended upon it. He may not have known the laws, he may not have known the words conduction, convection and radiation, but he did not need to know these concepts, because he knew them as a fact of experience. He could see that the smoke went upwards through the chimney, he could feel the heat radiating from the hot coals and from the glowing iron. His classroom was his life and his work.

Did he need to know the Clausius statement that heat always flows from hot to cold and never vice-versa, as he plunged his red-hot irons into a trough of water? No, he knew for a fact of experience that what is hot will certainly become cold in the course of time.

Now it is sometimes said that 'entropy' is a difficult concept. Really? It would seem to me that Gautama the Buddha understood this, thousands of years ago. After all it was entropy that set him off on his search for enlightenment. Certainly as a splendid and athletic young man as he saw before him old people and dying people, and as he pondered about old age and death, facts of life that had been shielded from him, certainly he would not have formulated anything about high entropy and low entropy. But he would have known these laws just the same, just as most of us know these laws. We know very well that an acorn has high entropy; that is because an acorn has the potential to become a mighty oak and to live in splendour for three or four hundred years. As I am now 81 I am increasingly aware that I have low entropy and no matter how hard I try to keep active in body and mind, however careful I am of my diet and exercise, this bodily mechanism in which my I is encased is decaying inevitably.

The birth of a baby is always greeted with joy – why? Because the baby has high entropy, has the potential to grow into manhood or womanhood. And if a young person is struck down too young, as many soldiers are, we grieve because they have not had the time to achieve their true potential.

Not everyone I know has even heard of the adiabatic lapse rate – however anyone who has gone up Snaefell in the Isle of Man, even by the mountain railway, or anyone who has climbed Goat Fell in the Isle of Arran, or climbed upwards anywhere on a mountain, anyone will know for a fact that as they go up it gets colder and the air gets more difficult to breathe. The Physicist may be able to explain why with altitude it gets colder, why there is snow on the tops of mountains, but nobody doubts the facts of their experience. As I have bathed in midsummer in the Lac de la Cavetaz at Sallanches in the Haute-Savoie, I have watched with delight the enormous snowcap on the mountain so rightly called Mt Blanc. How gloriously, gloriously shining white is Mt Blanc!

Do you imagine that the ancients did not know about TSI? Of course, they would not have called it Total Solar Irradiance, but those whom we somewhat naively term sun-worshipping pagans undoubtedly knew that the sun was the giver of light and warmth and the prime driver of climate. Indeed, if we were not beset by a barrage of misinformation, we would all know that the sun is just that – the driver of climate. Of course, the worship of the sun is not a thing of the past – indeed it has never been more prevalent than it is today. The tourist industry is largely based on the fact that the majority of Holy Day makers head for the sun if they possibly can.

Professor Brian Cox in his latest TV series 'Wonders of the Universe' assured us that the Arrow of Time can never be reversed. I wondered why he should make such a point of something that is seemingly self-evident. But put into context, not just with the motion of the Planets round the Sun, but also with the fact that the whole Galaxy of the Milky Way is revolving in stately and massive progress I began to see why Time is in fact a Law of Physics. It may take the Milky Way some 250 million years to complete an orbit, but what is certain is that not one iota can be reversed.

Once again one can say that this Law was grasped in its essence long, long ago by the Persian Poet, Omar Khayyam:

> **The Moving Finger writes; and, having writ,**
>
> **Moves on: nor all thy Piety nor Wit**
>
> **Shall lure it back to cancel half a Line,**
>
> **Nor all thy Tears wash out a Word of it.**

What the poet says corresponds to the facts of experience. Georgy Ivanovich Gurdjieff in his massive allegorical work "All and Everything" referred to the passage of Time as the Merciless Heropass.

So that when an esteemed colleague of mine declared that
*'the overarching issue is **climate change** and how to reverse it. We have a brief window of opportunity, and unless we act now the chances of avoiding a serious collapse are remote'* it means that he has not understood a fundamental Law of Physics. He can no more reverse changes of climate than he can reverse the revolutions of the Milky Way.

He was working on the assumption that man can and does control climate. This is a completely modern-day and atheistic heresy. I would even say blasphemy in the truest sense of the word.

What the Physicist is asserting, what the Physicist is examining are the Laws of Creation and the Laws of Destruction - Vishnu and Shiva in the Hindu pantheon. Unwittingly, even though he may think otherwise, the Physicist is a de facto theist.

Greenhouse Gases in the Atmosphere Cool the Earth!

Updated version – 10 May 2011

Robert Ashworth, Nasif Nahle and Hans Schreuder *
Introduction

In 2001, the United Nation's Intergovernmental Panel on Climate Change (IPCC) announced that carbon dioxide (CO_2) was causing the earth to warm and developed computer models to predict how much the earth would warm in the future. Does any empirical scientific evidence exist to support this premise of the IPCC? The answer is no, in fact it is just the opposite, CO_2 has a cooling effect.

The major components in the atmosphere that cause the earth to be cooler than it would be otherwise are the so-called greenhouse gases. Calling these gases greenhouse gases are misnomers. Yet, many meteorologists have blamed water vapor and carbon dioxide in the atmosphere for warming the earth. Below is an excerpt from a paper [1] written by meteorologists from the ***National Oceanic and Atmospheric Administration (NOAA)***. *"Water vapor plays the central role in the atmospheric branch of the global hydrologic cycle and is the most abundant greenhouse gas. <u>Climate models used for estimating effects of increases in greenhouse gases show substantial increases in water vapor as the globe warms and this increased moisture would further increase the warming</u>."* They got it backwards about water vapor just as Al Gore did about CO_2 in his "Inconvenient Truth" presentation of the Vostok Ice Core data.

Gore's "Inconvenient Truth" Documentary had Cause and Effect Reversed

In Al Gore's presentation of his "Inconvenient Truth" documentary, he conveniently separated the Vostok Ice core temperature and CO_2 graphs so you could not see which came first, a warming spike or a

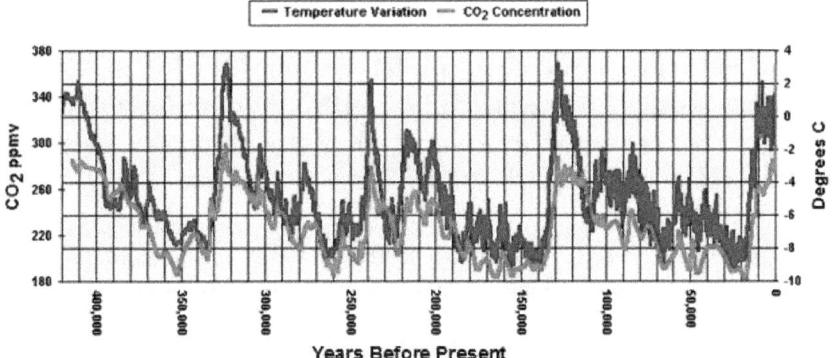

CO₂ spike. He said that a CO₂ spike came first but alas, it was the just the opposite as shown in the Vostok Ice Core graph in Figure 1.

Fig. 1. Vostok, Antarctica Ice Core Data [2].

When both lines are combined on one graph, suitably enlarged and viewed in the correct direction from left to right, it is clearly seen that a warming spike always comes first (blue line) followed by a CO₂ increase (red line) some 800 years later.

There are many reasons for the oceans to warm and cool over long periods of time and those influences are not the subject of this paper. Warming of the oceans reduces the solubility of CO₂ in water which results in the liberation of CO₂. An additional effect of the oceans warming up is that more water vaporizes and for each pound of water evaporated 1,000 Btu of cooling occurs. Increased water vapor and CO₂ in the atmosphere then causes a further cooling effect through the reflection of more radiation back to outer space. Nature has this under control.

Gore also gave no explanation what would cause a CO₂ spike to occur in the first place. What is so disturbing here is that, just like Al Gore, many climatologists and meteorologists seem to have a problem discerning cause and effect. It is very simple, if what you call an effect comes first, you have it backwards; a cause always comes first to produce the effect.

Proof Greenhouse Gases Cool the Earth
Proof 1: 9-11 Analysis

NASA scientists claimed Cirrus clouds, formed by contrails from aircraft engine exhaust, are capable of increasing average surface temperatures enough to account for the warming trend in the United States that occurred between 1975 and 1994. *"According to Patrick Minnis, a senior research scientist at NASA's Langley Research Center in Hampton, Va., there has been a one percent per decade increase in cirrus cloud cover over the United States, likely due to air traffic. Cirrus clouds exert a warming influence on the surface by allowing most of the sun's rays to pass through but then trapping some of the resulting heat emitted by the surface and lower atmosphere.*[3]*"*

This explanation is wrong. These clouds will cool the earth, not warm it. There is more radiant energy coming from the sun to the earth than from the earth to the sky. More radiant energy will be blocked during the day than will be blocked leaving the earth at night (insulating effect). The overall effect is cooling, not warming.

This cooling effect of water vapor was proved following the 9-11 terrorist attacks. Atmospheric scientists studied the effect of water vapor on temperature in the wake of the attacks. The Federal Aviation Administration (FAA) prohibited commercial aviation over the United States for three days following the attacks and this presented a unique opportunity to study the temperature of the earth without airplanes and their contrails.

Dr. David Travis, an atmospheric scientist at the University of Wisconsin, along with two other scientists, looked at how temperatures for those three days compared to other days when planes were flying. They analyzed maximum and minimum temperature data from about 4,000 weather stations throughout the conterminous (48 states) United States for the period 1971–2000, and compared those to the conditions that prevailed during the three-day aircraft grounding period and the three days when planes were flying before and after the grounding period. This research effort was sponsored by grants from the National Science Foundation.

They found that the average daily temperature range between highs and lows was 1.1 degrees C higher during September 11-14 (shown graphically in Figure 2) compared to September 8-11 and September 14-17 with normal air traffic.

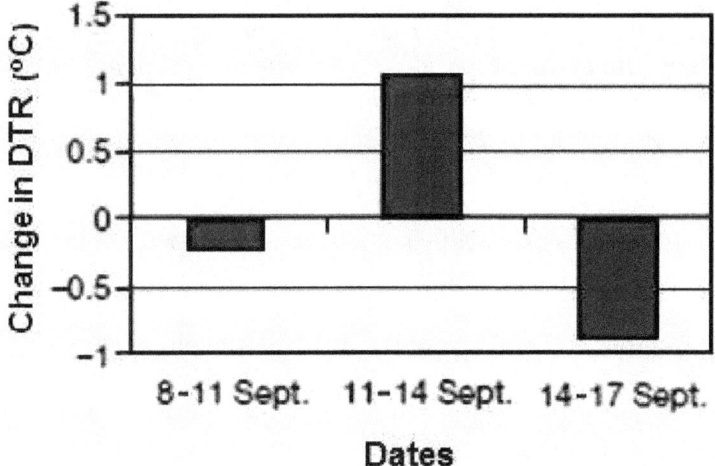

Fig. 2. Average diurnal (daily) temperature range (DTR)[4]

The data proved that contrails (condensed water vapor trails) have a net cooling effect. You cannot just look at a nighttime effect only, like the IPCC climatologists and meteorologists have done, both day and night must be included to determine the overall effect. Water vapor, CO_2 and particulates in the atmosphere all reflect as well as scatter some radiant energy back to outer space and this causes cooling.

Here is a simple test, go outside when the sun is shining, see how warm you feel when you are in the direct sunlight and compare that with how warm you feel when a cloud goes overhead and you are in the shade of the cloud. Of course you feel cooler in the shade of the cloud; a child knows this. So Dr. Travis confirmed this with scientific analysis of real data that most people on this planet already know.

Everyone also knows that cloud cover at night (more insulation) prevents the earth from cooling off as fast as it does when there are no clouds. However, on a relatively clear night if a cloud goes overhead you cannot feel any warming effect of the cloud, so this insulating effect is shown to be minimal compared to the daytime effect.

Another simple observation that shows Minnis and other warming supporters have it backwards. However, Minnis acknowledged the cooling effect found by Travis [5]. He and three other authors wrote, "*Instantaneously, contrail radiative forcing can warm the atmosphere and warm or cool the earth's surface, apparently reducing the diurnal range of surface temperature.*" This is like saying if you light a fire you may get warmer or cooler from radiative forcing.

Is this flawed logic why they have developed such illogical conclusions about the climate?

Proof 2: Comparison of Earth And Mars average temperatures

Both Earth and Mars rotate around the Sun and also rotate on their axes. The rotation time for Earth is 23.9 hours and for Mars is 24.6 hours so they are similar but Mars is only around 10% of the mass of the Earth. The atmospheric pressure on Earth is 1 atmosphere and the atmospheric pressure on Mars is 0.007 times the earth's atmosphere. The atmosphere on earth is primarily nitrogen (79%) and oxygen (21%) and on Mars it is approximately 95% carbon dioxide and 5% nitrogen. So the Mars atmosphere is a very small fraction of the Earths. There is approximately 400 ppmv of CO_2 in the Earth's atmosphere or (400/1000000) x 1 = 0.0004 atmospheres (earth). The CO_2 in the Mars' Atmosphere is 0.95 x 0.007 = 0.00665 atmospheres (earth). However if you add water vapor which they also call a greenhouse gas the partial pressure of CO_2 plus water vapor at say 1 vol.% in the earth's atmosphere the total partial pressure for the two is (0.01 + 0.0004) or 0.0104 compared to the 0.00665 for CO_2 on Mars, so if greenhouse gases caused warming, the earth's atmosphere would warm more than the Mars atmosphere when hit by radiant energy.

Is this what happens? No! The earth gets hit on average by 1367.5 watts/m^2 and Mars by 589.2 watts/m^2 of solar irradiance [6]. The average temperature on earth is 288.3K* and the average temperature on Mars is 208.3K [7]. Now then, if the Earth had the same composition (no free water) and atmosphere as Mars, based on the solar irradiance hitting it, compared to Mars, the average temperature on earth would be (1367.6/589.2) x 208.3 = 483.5K. The earth's atmosphere is shown by this analysis to have a cooling effect of (483.5K -288.3K) or 195°C more than the Mars atmosphere.

*K is the symbol for Kelvin, the absolute temperature scale used in scientific calculations and zero K is equivalent to minus 273.15°C. Convention has it that no degree symbol is used for K.

Does Atmospheric CO_2 Change Correlate with Earth Temperature Change?

Having conclusively proved that water vapor and clouds have a cooling effect, does a correlation of real data exist between the concentration of CO_2 in the atmosphere and the earth's temperature? No, that does not exist either, look at Figure 3 below developed by Joseph D'Aleo a certified meteorologist that was extended by the authors from 2008 through November 2010.

Fig. 3. Earth Temperature and Atmospheric CO_2 Concentration [8]. Even a non-scientist can see there is no correlation between CO_2 concentration in the atmosphere and the earth's temperature. The CO_2 has been on a continuous upward trend - not true for the earth's

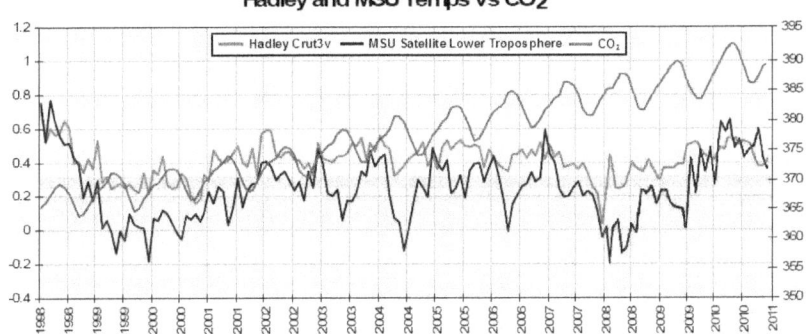

temperature. CO_2 cools the earth like water vapor does but since it is in parts per million in the atmosphere, unlike water vapor, the effect is so minimal it cannot be detected.

Atmospheric concentrations of CO_2 were taken at the Mauna Loa Observatory in Hawaii. Two sets of temperature measurements are included, one set by NASA's Microwave Sounding Unit (MSU) for the troposphere and the other by the UK's Hadley Climate Research

Unit for the land and sea. Both show normal temperature variations over time as CO_2 increased from 366 ppmv in January 1998 to 389 ppmv in November 2010. Note as well that the lower troposphere temperature (blue line) in November 2010 was some $0.37^{\circ}C$ lower than it was in the normally colder month of January in 1998, so overall there is no global warming trend anyway.

Man-made global warming advocates also say that CO_2 builds up in the atmosphere over a 50 to 250 year period, but this is not true. The graph above shows that the CO_2 concentration oscillates based on the growing seasons. The cycles of the CO_2 concentration swing is in the 5 to 8 ppmv range. If CO_2 stayed in the atmosphere for long periods before being consumed, the cyclical effect of the growing seasons would not be seen. It is clear that nature reacts very fast in its consumption of carbon dioxide. The steady rise in atmospheric CO_2 is most likely linked with the time period known as the Medieval Warm Period, some 800 years ago, as that would confirm the ice - core records where higher atmospheric CO_2 slowly follows higher ocean temperatures.

Figure 4 shows the likely reality of this warmer period with the unlikely hockey-stick illusion developed by Michael Mann through data manipulation and once used by the UN IPCC.

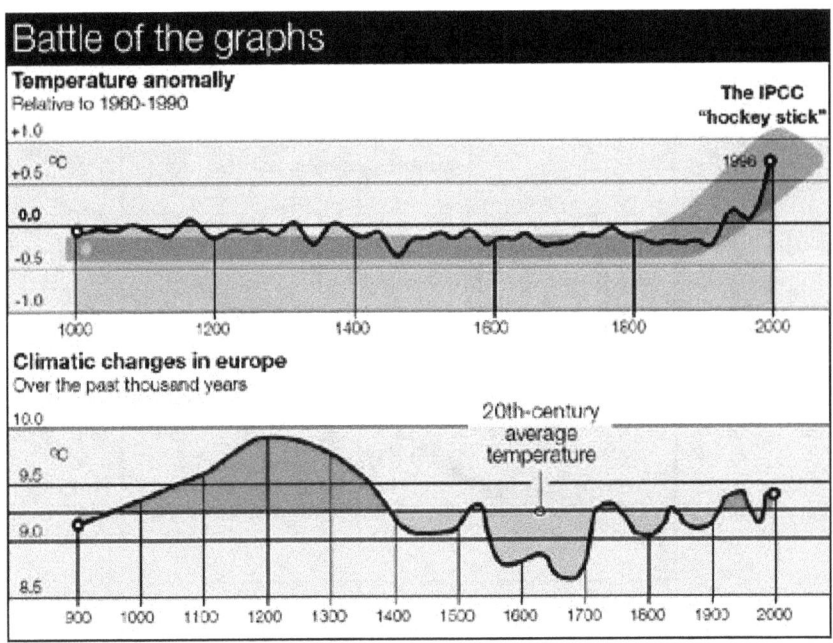

Fig. 4. Propaganda graphics versus reality [9].

Do UN IPCC Predictions fit in with Actual Observations?

The graph in Figure 5 shows the IPPC computer modeling projections from the year 2000 to 2100 based on various assumptions linked to increasing CO_2 in the atmosphere. The black line from 1998 to 2010 was added by the authors to the IPCC graph; it shows the actual measured surface temperatures.

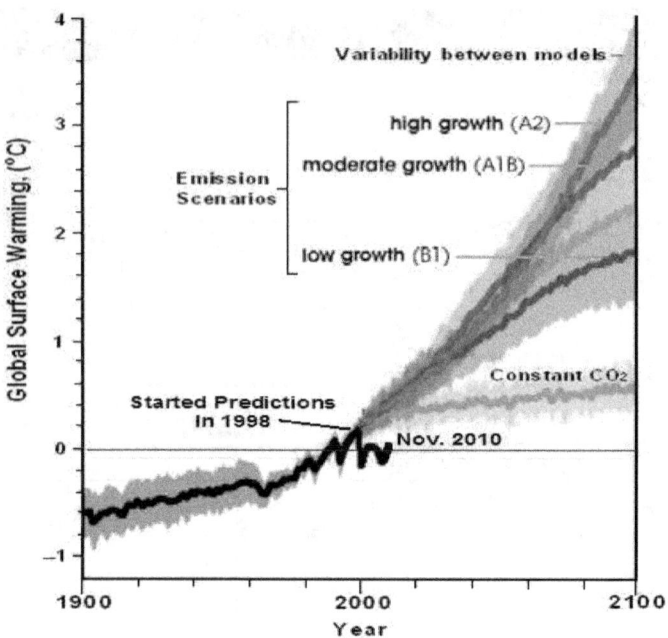

Fig. 5. IPPC Predictions compared to Actual Surface Temperature Measurements [10].

The actual temperature for November, 2010 was some 0.32 °C cooler than the IPCC projection based on the lowest assumed growth rate of CO_2 they used. On an actual temperature basis, one sees that the IPCC models predict temperatures that are not even close to actual measurements. Yet they continue to use these flawed computer models, while dismissing the actual temperature measurements by claiming they cover too short a time period. We will wait and see who is right.

It reminds one of the computer models used to predict where Hurricane Ike would hit the U. S. in 2008. Five meteorological models all predicted the hurricane would hit the west coast of Florida, then changed it to New Orleans, then to Galveston, down to Corpus Christi and then back up to Galveston, where it finally hit, all of this over a five day period. Here again, historically, most meteorological models have not proven trustworthy. These IPCC predictions are a result of a head-in-the-sand approach that produces the classic "garbage in - garbage out" analysis for computer models.

Human Made Emissions of Carbon Dioxide (CO_2)

You may not realize this, but CO_2 emissions created by human activities such as combustion of fuels, etc. (called anthropogenic emissions) are miniscule compared to the emissions of CO_2 from nature. Table 1 was developed by the UN IPCC. It shows annual CO_2 emissions to the atmosphere from both nature and human-made sources and how much of the CO_2 emitted is re-absorbed by nature.

TABLE 1. GLOBAL SOURCES AND ABSORPTION OF CO_2

Carbon Dioxide:	Natural	Human-Made	Total	Absorption
Annual (Million Metric Tons)	770,000	23,100	793,100	781,400
% of Total	97.1%	2.9%	100%	98.5%

Source: Intergovernmental Panel on Climate Change, Climate Change 2001: The Scientific Basis
(Cambridge, UK Cambridge University Press, 2001), Figure 3.1, p. 188.

So Nature absorbs 98.5% of the CO_2 that is emitted by nature and humans. There were many claims about how long man-made CO_2 remained in the atmosphere in the IPCC reports and they varied from a few years to as much as a 100 years in the political summary of the 2007 IPPC report and even more wild claims outside of the IPCC reports mentioned more than a thousand years. Let us fill a container with 1 unit of CO_2 generated at a uniform rate in a year. Now we will drain that container at a rate that removes 98.5% of it in a year, but we will keep removing it at that rate until the container is empty.

The time to empty the container is (1/0.985) years or 1.015 years. Seeing that 1.015 years is 370.73 days - with 365.25 days per year – and knowing that the drain rate was constant, this means that a half unit of CO_2 was drained in 370.73 / 2 = 185.36 days. This rate of removal calculation is pretty simple compared to the gyrations people go through to get the many years numbers, which makes this number much more believable.

Using the table above in combination with the average concentration of 373 ppmv of CO_2 seen in the atmosphere in 2001, one sees that the CO_2 caused by all of our activities amounts to only 10.8 ppmv for 2001. If, on a worldwide basis, we eliminated all anthropogenic CO_2 emissions in November 2010 (see Figure 3), we would go back to the level we had in 2004, making carbon dioxide emission reductions counter-productive and a complete waste of resources.

As CO_2 increases in the atmosphere, nature's controlling mechanism causes plant growth to increase via photosynthesis; CO_2 is absorbed, and oxygen is liberated. Photosynthesis is an endothermic (cooling) reaction. Furthermore, a doubling of CO_2 will increase the photosynthesis rate by 30 to 100%, depending on other environmental conditions such as temperature and available moisture [11].

More CO_2 is absorbed by the plants due to the increased concentration of CO_2 in the atmosphere available for conversion to carbohydrates. Nature therefore has in place a built-in mechanism to regulate the CO_2 concentration in the atmosphere that will always completely dwarf our feeble attempts to regulate it. Even the US Dept of Energy knows this, yet gets it all wrong when they write a paper on it, see page 13 and reference 28.

Further, no regulation by us is necessary because CO_2 is not a pollutant; it is part of the animal and plant life cycle. Without it, there would be no life on earth. Increased CO_2 in the atmosphere increases plant growth, which is a good thing during continued world population growth and an increasing demand for food

No Greenhouse Signature in the Atmosphere

The IPCC developed a theoretical greenhouse signature through a computer model. Their theoretical greenhouse signature is very distinct - see Figure 6. If this signature were present, warming would be concentrated in a distinct "hot spot" about 8 to 12 km up over the tropics, with less warming further away, turning to cooling above 18 km.

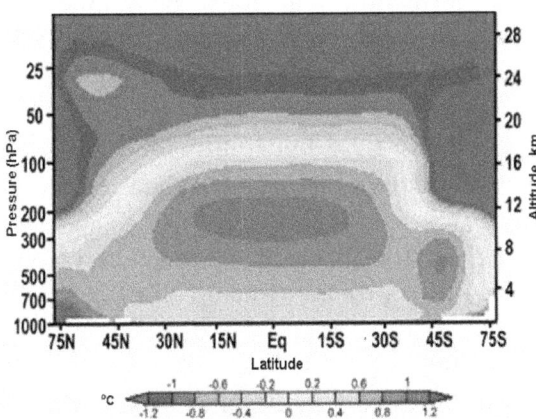

Fig. 6. Theoretical Greenhouse Signature

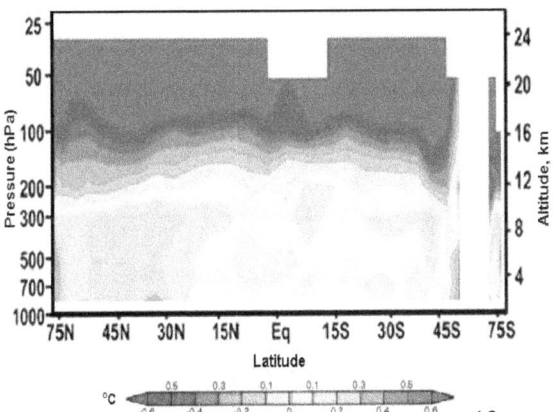

Fig. 7. Actual Observed Signature [13].
(UN climate models) [12].

Actual measurements have been taken where the warming should be occurring according to the models using satellites and balloons. The real observed signature is shown in Figure 7. As one can clearly see, the predicted IPCC "Greenhouse" signature is not seen - no "hot spot" exists!

Heat Transfer by Radiation from Gas to Solid

The Stefan-Boltzmann Law, that has been verified experimentally, states that the total radiation from a black body is proportional to the fourth power of its absolute temperature. For non black bodies there is an emissivity correction that must be applied based on the emissivity of the non black body compared to a black body at the same temperature.

Radiation between gases does not follow the Stefan-Boltzmann law for radiation between solid surfaces[14]. When radiant energy passes through certain gases and vapors an appreciable portion within certain frequencies may be absorbed but practically no radiant energy will be absorbed within other frequencies.

The following method of estimating the rate of heat transfer by radiation from carbon dioxide and water vapor to a bounding solid surface was developed by Hottel and Egbert [14].

The rate of heat transfer by radiation between a hemispherical mass of gas with a radius (L), partial pressure of gas emitting radiation of (p) atmospheres, at the uniform absolute temperature of the gas (T_g), and a small element of solid surface at its absolute temperature (T_s) with a surface emissivity (e_s) located on the base of the hemisphere at its center is:

$$Q/A = 0.173 \; e_s * [e_{gg} *(T_g/100)^4 - e_{gs}*(T_s/100)^4]$$

Where, Q = Rate of heat transfer between gas and
solid A = Area of heat absorbing surface
e_{gg} = emissivity of the gas at T_g
e_{gs} = emissivity of the gas at T_s

The emissivity of the gas depends on the temperature of the gas and the product pL. When the gas contains a mixture of CO_2 and water vapor, the combined radiation from these two gases is less than the sum of their separate effects and a correction factor is applied based on the partial pressures of the two gases.

One can see from this equation that the net heat transfer is always from the hotter to cooler body, never vice versa. So one cannot show a balance where a cooler body is heating a warmer body; that is impossible and is a violation of the second law of thermodynamics!

Many scientists are confused over thermal equilibrium issues associated with analyzing the climate. There are two fundamental errors on how many handle the physical concepts of the Earth's thermal equilibrium and of blackbodies. The Earth cannot reach thermal equilibrium for many reasons, but mainly because the Earth is not a blackbody because, simply, it does not absorb all the energy it receives [15]. For the same reason, the atmosphere is not, and does not behave like, a blackbody and it is not even a near-blackbody system.

Let us examine the first mistake. Thermal equilibrium refers to the approaching to almost the same temperature of two interrelated systems, where one of them is warmer than the other system and the warmer system transfers some of its energy to the colder system [16]. In other words, between two interrelated systems, the warmer system does work, in the form of heat, on the colder system, not the other way around [16, 17].

On planet Earth, which has a great diversity of subsystems [18], it is physically impossible to obtain thermal equilibrium spontaneously, i.e. in nature. Simply, let us measure the energy contained by two or more small adjacent volumes of air; one will offer an energy density higher or lower than the other adjacent air volumes. This is considering solely one of those subsystems, the atmosphere.

The Earth rotates over its own axis, giving place to the alternating diurnal and nocturnal periods. We cannot speak of a planetary thermal equilibrium when a half of the terrestrial sphere receives solar light, gaining energy incoming from the Sun, whereas the other is in darkness, losing the solar energy that it had received during the daytime. We always find differences of the density of energy between one hemisphere and the other.

Besides, the thermal equilibrium concept only happens when the temperature of both systems keeps constant [17]. When have we seen the temperature of the Earth being constant? The answer is nowhere and never. The temperature suffers fluctuations many times a second all day long. Whichever subsystem we examine, even the temperature of the endothermic living beings, which self-regulate their body temperature, fluctuates many times a second [19].

On the other hand, if AGW hypothesis proponents are referring to a thermal equilibrium between the density of the incident solar energy on the Earth's surface and the Earth's surface density of energy, they are making a most important mistake because the Earth's surface never reaches the density of energy that the incoming stream from the Sun has. The density of energy of the solar stream is always higher than the density of energy of the surface. The absorbent system disperses the energy that it does not "use", like work, towards other systems; to outer space, for example. We know this class of energy as "heat", i.e. energy in transit.

Obviously, the energy in transit, or in the very moment of the transferring process, by any means, cannot be stored; it is energy in transit from one energized system to another less-energized system.

As a result, thermal equilibrium only occurs in mathematically isolated systems where the primary source of energy for both systems, the emitter and the absorber, has been artificially eliminated. The real Earth system cannot ever reach a thermal equilibrium because its primary source of energy is always present and the Earth rotates. The Sun is this source in this case and, as we have observed in the last decades, it is not too stable. With the latter assertion, it is not being said that the Sun is a meta-stable system, not even in the long term and this statement must be clear. Have we ever seen the global climate as being stable or, at least, quasi-stable? No, we have never seen such a phenomenon.

Regarding the concept of blackbody systems, the actual physical concept says that a blackbody is an idealized system that "absorbs" all the energy, at all spectral bands and wavelengths and all temperatures that it receives from a source [20], which, consequently must be in a higher energy density level than the absorbent blackbody. We cannot say the atmosphere is a blackbody because it has a very limited (low) absorptivity.

By applying the scientifically accepted algorithms, which derive from experimentation [16], to calculate the absorptivity of the air, we find that it is 0.01, at the best composition of the air, i.e. mixed with dust particles and 4% of water vapor. Could anyone of us argue that the atmosphere is a near-blackbody system? No, for the only case of a near-blackbody system, but not one of those idealized systems, is moistened clay with organic matter that has an absorptivity of 0.95 [20] or water vapor, with an absorptivity of 0.75 [20].

Carbon dioxide, on its own, has an absorptivity of 0.002; in the overall panorama, Oxygen, with its 0.007 of absorptivity/emissivity, is 3.5 times a better absorber/emitter than carbon dioxide. For that reason alone, carbon dioxide *is not* a blackbody or even a "near-blackbody" system.

Greenhouse Effect is Impossible

The IPCC adopted the work completed by Kiehl and Trenberth of the National Center for Atmospheric Research, Boulder Colorado to show how radiative forcing from greenhouse gases causes the earth to warm [21]. Here is a statement from that paper:

> The long wave radiative forcing of the climate system for both clear [125 W/m^2 (watts/square meter)] and cloudy (155 W/m^2) conditions are discussed. We find that for the clear sky case the contribution due to water vapor to the total long wave radiative forcing is 75 W/m^2, while for carbon dioxide it is 32 W/m^2.

Really, when the water vapor concentration in the atmosphere averages around 1 volume % (or 10,000 ppmv) and carbon dioxide concentration is less than 400 ppmv? The CO_2 concentration is only 4% of the water vapor concentration. In the Hottel and Egbert correlation the only difference between water vapor and carbon dioxide regarding the radiation effect is their partial pressures. Partial pressures of gases are proportional to their volumetric concentrations. Based on this and using the water vapor effect as a basis at 75 W/m^2 then the CO_2 effect would be 3 W/m^2, not the 32 W/m^2 stated. Figure 8 shows an updated graph from the original work completed by Kiehl and Trenberth in 1997.

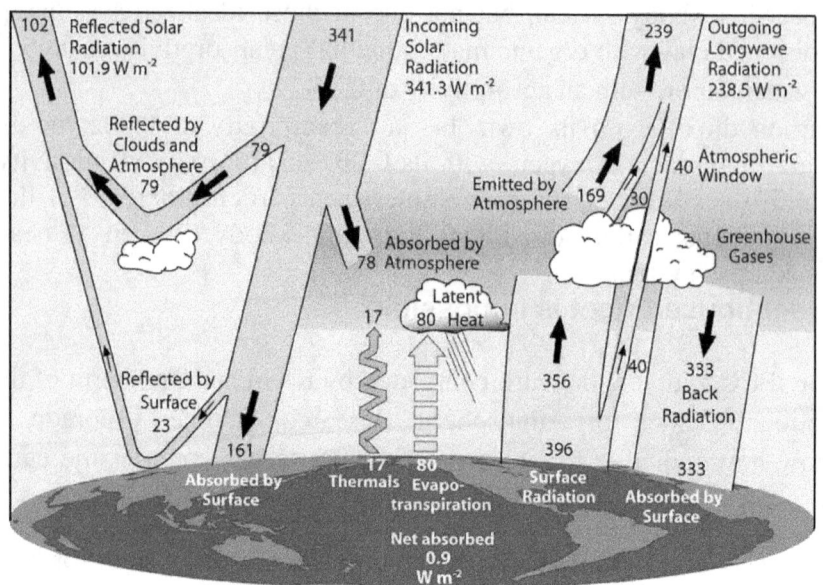

Figure 8. The Global Annual Mean Earth's Energy Budget for March 2000 to May 2004 [22].

The first problem seen in this graph is a serious violation of the First Law of Thermodynamics. Completing a balance only on the energy entering and leaving the earth, one sees on the middle to left side of the graph that 161 W/m^2 hits the earth from the sun. The earth then loses 97.9 W/m^2 to thermal and evaporation/transpiration losses with 0.9 W/m^2 being retained by the earth. This leaves only 63.1 W/m^2 not accounted for. However, on the right side of the earth it shows 396 W/m^2 leaving the earth. If you only have 63.1 W/m^2 available, no more than that can leave the earth. Well you say an additional 333 W/m^2 in back radiation from the clouds hits the earth and this then provides the balance. Mathematically you can do anything you like of course, as long as input equals output, but where is the reality in such a computation?

The only problem here is, where did this added energy come from? It came from the bogus 396 W/m^2 shown radiating from the earth's surface to the clouds. If what was presented here were true, for every unit of energy in, one would get back an additional (396 ÷ 63.1) = 6.28 units of energy. The First Law of Thermodynamics states that energy can be transformed from one form to another but cannot be created nor destroyed. Therefore, the graph violates the First Law of Thermodynamics.

It also violates the Second Law of Thermodynamics by showing energy radiating from cooler clouds back to a warmer earth. One version of the second law states no process is possible where the sole result is the transfer of heat from a body of lower temperature to a body of higher temperature. Maybe they are still counting on the non-existent hot spot in the greenhouse signature to accomplish this. Real data shows the IPCC computer models, both for predicting the effect of CO_2 on the earth's temperature and the climate signature, are misconstrued.

As a final rebuttal of the influence of carbon dioxide over the climate, the alleged IPCC greenhouse effect is a non-existent effect. No greenhouse, whether made from glass, plastic, cardboard or steel will reach a higher inside temperature due to the magic of re-radiated IR energy. If it did, engineers would have long ago been able to design power stations made from air, mirrors and glass, extracting more energy out of it than was put into it - if only!

Is the Climate Change Agenda a Fraud?

There is supporting evidence that indicates that the Climate Change agenda is and always has been just a fraud [23]. Why do we call it a fraud? An event now referred to as Climategate publicly began on November 19, 2009, when a whistle-blower leaked thousands of emails and documents central to a Freedom of Information request placed with the Climatic Research Unit of the University of East Anglia in the United Kingdom. This institution had played a central role in the "climate change" debate: its scientists, together with their international colleagues, quite literally put the "warming" into Global Warming: they were responsible for analyzing and collating the measurements of temperature from around the globe from the present to the distant past.

Dr. John Costello relays[23], "Climategate has shattered that myth (*the myth of global warming*). It gives us a peephole into the work of the scientists investigating possibly the most important issue ever to face mankind. Instead of seeing large collaborations of meticulous, careful, critical scientists, we instead see a small team abusing almost every aspect of the framework of science to build a fortress around their 'old boys club', to prevent real scientists from seeing the shambles of their research. Most people are aghast that this could have happened; and it is only because climate science exploded from a relatively tiny corner of academia into a hugely funded industry in a matter of mere years that the perpetrators were able to get away with it for so long."

Stephen McIntyre, a University of Toronto mathematics graduate first questioned the accuracy of the "hockey stick" temperature graph mentioned previously. He wondered how Michael Mann, head of Penn State's Earth System Science Centre, reconstructed the temperatures to produce such a detailed graph. McIntyre then sent an e-mail to Michael Mann in the spring of 2003 [24], asking him for the location of the data used in his study. "Mann replied that he had forgotten the location," he said. "However, he said that he would ask his colleague Scott Rutherford to locate the data. Rutherford then said that the information did not exist in any one location, but that he would assemble it for me. McIntyre thought this was bizarre and so do we.

Here is what was later found concerning the temperature data Mann used. Five organizations publish global temperature data [25]. Two – Remote Sensing Systems (RSS) and the University of Alabama at Huntsville (UAH) – are satellite datasets. The three terrestrial institutions – NOAA's National Climatic Data Center (NCDC), NASA's Goddard Institute for Space Studies (GISS), and the University of East Anglia's Climatic Research Unit (CRU) – all depend on data supplied by ground stations via NOAA.

Around 1990, NOAA began weeding out more than three-quarters of the climate measuring stations around the world. It can be shown that systematically and purposefully, country-by-country, they removed higher-latitude, higher-altitude and rural locations, all of which had a tendency to be cooler. The thermometers kept were near the tropics, the sea, and airports near bigger cities. These data were then used to determine the global average temperature and to initialize climate models. From 1960 through 1980, there were more than 6000 stations providing temperature information. The NOAA reduced these to fewer than 1500. Calculating the average temperatures this way ensured that the mean global surface temperature for each month and year would show a false-positive temperature anomaly, a bogus warming trend. Interestingly, the very same stations that were deleted from the world climate network were retained for computing the average-temperature base periods, further increasing the bias towards warming.

An internal study of the U.S. EPA[26] completed by Dr. Alan Carlin and John Davidson concluded the IPCC was wrong about global warming. One statement in the executive summary stated that a 2009 paper [27] found that the crucial assumption in the Greenhouse Climate Models (GCM) used by the IPCC concerning a strong positive feedback from water vapor is not supported by empirical evidence and that the feedback is actually negative. This is exactly what we have shown here, water vapor in the atmosphere causes a cooling effect, not a warming one.

EPA tried to bury the report. An email from Al McGartland, Office Director of EPA's National Center for Environmental Economics (NCEE), to Dr. Alan Carlin, Senior Operations Research Analyst at NCEE, forbade him from speaking to anyone outside NCEE on endangerment issues. In a March 17 email from McGartland to Carlin, stated that he will not forward Carlin's study. *"The time for such discussion of fundamental issues has passed for this round. The administrator (Lisa Jackson) and the administration have decided to move forward on endangerment, and your comments do not help the legal or policy case for this decision. .... I can only see one impact of your comments given where we are in the process, and that would be a very negative impact on our office."* A second email from McGartland stated *"I don't want you to spend any additional EPA time on climate change."*

McGartland's emails demonstrate that he was rejecting Dr. Carlin's study because its conclusions ran counter to the EPA's current position. Yet this study had its basis in three prior reports by Carlin (two in 2007 and one in 2008) that were accepted. Another government cover-up, just what the United States does not need.

The US Dept of Energy knows what is going on: "Based on preliminary data from satellite measurements, Ramanathan and his colleagues concluded that clouds appear to cool Earth's climate, possibly offsetting the atmospheric greenhouse effect." [28] *Yet they continue to make basic errors, even stating that* "The decay time or residence time for carbon dioxide ranges from around 150 to about 500 years. Without carbon dioxide absorbers (sinks), the carbon dioxide residence time would be only 3.9 years, which is precisely its exchange rate." *What they ought to state is that with carbon dioxide absorbers (sinks), the carbon dioxide residence time is only 3.9 years.* They know the truth but can't say it, why?

Further, Dr. Noor van Andel in January 2011 updated his paper, *CO_2 and Climate Change*,[29] and explains in detail how climate scientists adjusted radiosonde (weather balloon) data to try to bring it into agreement with their computer models to show greenhouse gas induced global warming. This is quite the opposite of the normal scientific procedure of adjusting the models to fit the data. The unadjusted data does not show the elusive "hot spot" (greenhouse gas signature) predicted by climate models and conventional 'greenhouse' theory.

Most of the U.S. House of representatives agree with the fraud assessment. On February 19, 2011 they voted to eliminate U.S. funding for the Intergovernmental Panel on Climate Change. With a vote of 244-179, they said that it no longer wishes to have the IPCC prepare its comprehensive international climate science assessments.

The amendment was sponsored by Rep. Blaine Luetkemeyer (R-Missouri). He said;

> "The IPCC scientists manipulated climate data, suppressed legitimate arguments in peer-reviewed journals, and researchers were asked to destroy emails, so that a small number of climate alarmists could continue to advance their environmental agenda. Since then, more than 700 acclaimed international scientists have challenged the claims made by the IPCC, in this comprehensive 740-page report. These scientists represent some of the most respected institutions at home and around the world, including the U.S. Departments of Energy

and Defense, U.S. Air Force and Navy, and even the Environmental Protection Agency".

Conclusion

Based on empirical data, CO_2 causing global warming is clearly a figment of the UN IPCC's imagination. The only actual physical attribute that can be ascribed to atmospheric carbon dioxide is one of increasing the cooling efficiency of the total atmosphere, as recently demonstrated by Dr Noor van Andel [29]. For that matter, as discussed, all greenhouse gases, including the most notorious water vapor and carbon dioxide, cool the planet. Taxing carbon will have a deleterious effect on the economies of the world.

Many scientists, including the authors, see global warming from CO_2 as a cruel global hoax.

*

Robert Ashworth is a chemical engineer and member of the American Geophysical Union. He holds 16 US patents and some foreign patents on coal conversion processes and environmental control technologies. In 2001 the Kentucky Governor Paul Patton commissioned him a Kentucky Colonel for his work on clean coal technology.

Nasif Nahle is a University Professor and Director of the Scientific Research Division at Biology Cabinet, Mexico; a Biologist, 1st. Level Certificate in Scientific ICAM Research by the University of Harvard and certificated in Physics, Climatology, Meteorology, Paleontology and Geology. He's a member of the American Association for the Advancement of Science and the American Physical Society.

Hans Schreuder is a retired analytical chemist and now webmaster of the I Love My Carbon Dioxide website where he has been promoting scientific scepticism about the man-made aspect of climate change. He is also a co-author of the world's first book debunking the climate change hype based on the imaginary "greenhouse effect" – Slaying the Sky Dragon: Death of the Greenhouse Gas Theory".

References:

1. Ross, R. J., and Elliott, W.P., "Radiosonde-Based Northern Hemisphere Tropospheric Water Vapor Trends"
   Journal of Climate, Vol. 14, 1602-1612, July 7, 2000.

2. Petit, J.R., et. al., "Climate and Atmospheric History of the past 420,000 years from the Vostok Ice Core, Antarctica",
   Nature 399: 429-436, June 3, 1999.

3. Minnis, P., "Clouds Caused by Aircraft Exhaust May Warm the U. S. Climate", NASA Release 04-140, April 27, 2004.

4. Travis, D., A. Carleton, and R. Lauritsen, 2002: Contrails reduce daily temperature range. Nature, **418,** 601.

5. Minnis, P., J. K. Ayers, R. Palikonda, and D. N. Phan, 2004: Contrails, cirrus trends, and climate. J. Climate, 17, 1671-1685.

6. http://nssdc.gsfc.nasa.gov/planetary/factsheet/marsfact.html
7. http://nssdc.gsfc.nasa.gov/planetary/factsheet/planet_table_british.html

8. D'Aleo, J. S., "Correlation Last Decade and This Century CO_2 and Global Temperatures Not There"
   http://icecap.us/images/uploads/Correlation_Last_Decade.pdf. The authors updated the graph to extend from the original 2008
   end period until the end of November, 2010 using CO_2 data from the Mauna Loa Observatory and temperature data from
   NASA's Microwave Sounding Unit (MSU) for the troposphere and the other from the UK's Hadley Climate Research Unit
   for the land and sea; the data sources use are the same as that used by D'Aleo in the original graph.

9. http://ncwatch.typepad.com/media/2010/03/sierra-medieval-warm-period-evidence.html

10. NASA Earth Observatory, based on IPCC Fourth Assessment Report (2007)
    (http://epa.gov/climatechange/science/futuretc.html) and Hadley Climate Research Unit, Global Temperature Record,
    Phil Jones, http://www.cru.uea.ac.uk/cru/info/warming/

11. Pearch, R.W. and Bjorkman, O., "Physiological effects", in Lemon, E.R. (ed.), CO_2 and Plants: The Response
    of Plants to Rising Levels of Atmospheric CO2 (Boulder, Colorado: Westview Press, 1983), pp 65-105

12. Intergovernmental Panel on Climate Change (IPCC), 2007, p. 675, based on Santer et al, 2003. See also
    IPCC, 2007, Appendix 9C). Authors added actual temperature data from 1998 to Nov. 2010.

13. David Evans, "Carbon Emissions Don't Cause Global Warming", November 28, 2007,
    http://icecap.us/images/uploads/Evans-CO2DoesNotCauseGW.pdf

14. Hottel and Egbert, Trans. American Society of Mechanical Engineers, Vol. 63, p. 297, 1941

15. Wolfe, Robert W. and Wolfe, Mary Jane. *Taking Earth's Temperature-Blackbody Earth*.
    https://www.math.duke.edu/education/prep02/teams/prep-15/index.html

16. Modest, Michael F. *Radiative Heat Transfer-Second Edition*. 2003. Elsevier Science, USA and Academic Press, UK.

17. Thomas Engel and Philip Reid. *Thermodynamics, Statistical, Thermodynamics & Kinetics*. 2006. Pearson Education, Inc.
    Pp. 13, 16, 355

18. Peixoto, José P., Oort, Abraham H. 1992. *Physics of Climate*. Springer-Verlag New York Inc. New York. Page 8.

19. Mader, Sylvia S. *Human Biology*. 2004. The McGraw-Hill Companies Inc. New York.

20. Manrique, José Ángel V. *Transferencia de Calor*. 2002. Oxford University Press. England

21. J. T. Kiehl and Kevin E. Trenberth, Earth's Annual Global Mean Energy Budget, Bulletin of the American
    Meteorological Society, Vol. 78, No. 2, page 206, February 1997 (adopted by IPCC 2007)

22. Trenberth, K. E., Fasullo, J. T. and Kiehl, J., "Earth's global energy budget", Bulletin of the American
    Meteorological Society, Vol. 90, 311–323, July 2008

23. Costella, J.P., "Climategate Analysis", http://assassinationscience.com/climategate/

24. Marcel Crok, "Breaking the hockey stick",
    http://climaterealists.com/index.php?id=1642

25. D'Aleo, J. and Watts A., "Surface Temperature Records: Policy of Deception?"
    http://scienceandpublicpolicy.org/originals/policy_driven_deception.html

26. Carlin, A. and Davidson, J, "Proposed NCEE Comments on Draft technical Support Document for Endangerment Analysis for Greenhouse Gas Emissions under the Clean Air Act", March 9, 2009.
    http://cei.org/cei_files/fm/active/0/DOC062509-004.pdf

27. Gregory, Ken 2009, Climate Changing Science,
   http://www.friends of science.org/assets/documents/FOS%20Essay/Climate_Change_science.html

28. Alternatives to Traditional Transportation Fuels 1994, Volume 2, Greenhouse Gas Emissions, Appendix A
   The Chemistry and Physics of Global Warming: An Overview
   http://www.eia.doe.gov/cneaf/alternate/page/environment/appd_a.html

29. Andel, Noor van, Climate changes are not caused by greenhouse gases
   http://hockeyschtick.blogspot.com/2011/01/scientist-climate-changes-are-not.html

The Under-Whelming Evidence

Anthony Bright-Paul

Tuesday, August 02, 2011

Saturday morning I was watching the BBC NewsWatch programme with Raymond Snoddy, a Science Correspondent, defending Professor Steve Jones, who spoke of the 'overwhelming evidence of Global Warming.' This aroused my curiosity. Had I missed something? What was this overwhelming evidence? And why had not I been overwhelmed?

Professor Steve Jones, defending the impartiality of the BBC also said that one area of concern related to balance, on which issue he recommended that the BBC must make a distinction between well-established fact and opinion to avoid giving free publicity to marginal opinions.

That is a very curious statement from a famous geneticist. I like the conjunction between 'well-established fact' and 'opinion'. It is perfectly true that there is a large body of opinion that there is such a thing as Anthropogenic Global Warming, but where, pray, are the facts? Where is the evidence?

With atmospheric Carbon Dioxide increasing, there has been no increase in Global Warming, that is, taking the figures of the Climatic Research Unit. And since those figures are to be doubted by many since Climategate, they are totally unreliable anyway. The number of climate stations has fallen dramatically, and there are no climate stations on the 71% of the Earths' surface that is water. So there appears to be no warming to speak of in the last 10 or 12 years, in spite of all the brouhaha.

There is supposed to be a connection between Carbon Dioxide and this not-evident Global Warming. And the connection rests on one single attribute of Carbon Dioxide, the gas, that is '…Carbon dioxide is a greenhouse gas as it transmits visible light but absorbs strongly in the infrared and near-infrared.' (from Wikipedia.)

Much has been made of the fact that it absorbs strongly in the infra-red and near infra-red. However, '...**Thermal radiation** is electromagnetic radiation generated by the thermal motion of charged particles in matter. All matter with a temperature greater than absolute zero emits thermal radiation.'

All matter? Even a park bench?

On the basis that CO2 absorbs and then re-emits infra-red, a whole scare story has been built up, to the effect that man-made Carbon Dioxide in the atmosphere is causing this famous Global Warming, of which there is no overwhelming evidence. Indeed of which there is no evidence at all!

Indeed most of those scientists who support this view, including Al Gore, know very well that it is wrong. And it is wrong because the atmosphere, - the troposphere - is already freezing from some 7,500 feet and upwards. By 30,000 feet it is an ice-box at minus 45 degrees centigrade. He knows that very well. Do you imagine that all those alarmist scientists don't know that there is a continuum to Outer Space? All those diagrams by Professor Kevin Trenberth are surely flawed.

We have yet more nonsense, in so far that Carbon Dioxide is supposed to 'trap' heat. Really? How come then that Carbon Dioxide can be solidified to make Dry Ice, which is colder than Water Ice?

No. Carbon Dioxide can be trapped in a room with the windows all shut. It can be trapped in a car. It can easily be trapped in a kid's balloon. Gases can be trapped on purpose in those large balloons we see sometimes wandering in the sky, blown by the wind. Gases can be trapped in containers, in gas-holders. Gases can be compressed, and even liquified for ease of transport.

But can a Gas trap anything? Anything whatsoever? Try it. Just try lassooing something with a gas. Choose any gas you like, Oxygen, Nitrogen, Methane, Carbon Monoxide. Let me know of your success.

In the days of Copernicus, virtually everybody believed that the Earth was the centre of the Universe. Luckily for Copernicus he died just as his great book was published, where he averred that the Earth orbits round the Sun. So he escaped being burned at the stake like Giordano Bruno, who went farther than Copernicus in declaring that the Sun was a star and who compounded his error by also asserting that Jesus was a Prophet of God and not God Himself.

As for Galileo, who had a horror of torture, he was hounded by the Inquisition who tried to trap him into making some heretical statements, while we know now that he fully supported the Copernican thesis that the Earth moved round the Sun, thus showing that the Earth was not the centre of the Universe.

Reading the history of those times it is remarkable to see that the Lutherans, the Protestants who had broken away from the Church of Rome, were nevertheless just as intolerant as the Roman Church and equally ignorant.

So we come to the Law of the Octave as enunciated by Georgy Ivanovich Gudjieff, that everything in time becomes its own opposite. So we see that the religion of Jesus Christ that taught the Love of God and Surrender to the Will of God, (Thy Will be done) had within a few hundred years become a religion of extreme ignorance and intolerance. Where in the scriptures can anywhere be found anything that justified throttling a man gradually with ropes, or burning a man to death for his beliefs?

Such mediaeval intolerance would seem to be a thing of the past. In hindsight it is easy to see that the Churches had become corrupt political bodies, far removed from anything that can be truly called Religion. But in fact we have today an even more insidious and intolerant false religion, which attempts in every way to muzzle Science, and to distort the conclusions that must follow from a correct appreciation of the facts.

And what pray is Science? Anyone who has learnt Latin knows the principal parts of the verb: Scio, scire, scivi, scitum. Science is knowledge. Far from the BBC being impartial, the BBC has resolutely set its face against those scientists who aver that there is no Greenhouse Effect, that there is no man-made Global Warming, that any such warming as does occur is entirely through natural causes, and any changes in climate are likewise caused by Great Nature. Far from being impartial, no discussion, no debate, is allowed. Not since the 8th of March 2007 when Channel 4 premiered the documentary 'The Great Global Warming Swindle' has there been any meaningful discussion of this subject on TV, least of all by the BBC.

Now we have Professor Steve Jones actively proposing that minority opinions should not be heard at the BBC. But, were they ever? We know very well that if the Sceptics had the opportunity to debate openly through the 'impartial' BBC they would trounce their opponents in open debate.

This malaise is world-wide, as only this morning I heard from Australia the attempts of that false religion to prevent the Press there from even allowing a dissenting voice. In Canada a well-known Sceptic is being forced into litigation, effectively having the thumbscrews forced upon him for the great sin of speaking the truth.

How far have we come from the Middle Ages? Not very far. The methods are slightly different, the howling ignorance is the same and the Inquisition wears a Green cloak.

Inverse Square Law, General

From: John Etherington
Sent: Friday, September 16, 2011 4:29 PM

Any point source which spreads its influence equally in all directions without a limit to its range will obey the inverse square law. This comes from strictly geometrical considerations. The intensity of the influence at any given radius r is the source strength divided by the area of the sphere. Being strictly geometric in its origin, the inverse square law applies to diverse phenomena. Point sources of gravitational force, electric field, light, sound or radiation obey the inverse square law. It is a subject of continuing debate with a source such as a skunk on top of a flag pole; will it's smell drop off according to the inverse square law?

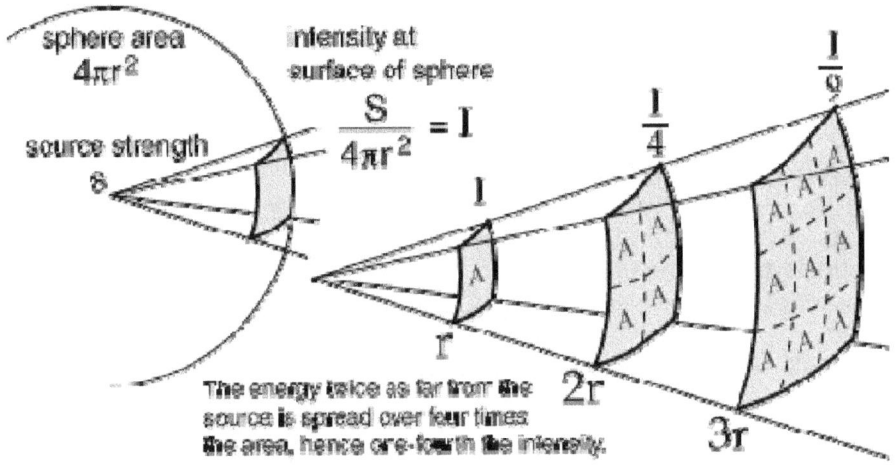

From: John Etherington
Sent: Friday, September 16, 2011 4:29 PM

Get your Heads round this!

05.11.2011

This exposition by Hans Schreuder is so magnificent that I must share it with you.

Hans Schreuder Exposition:-

Dear Tony,

The sun does not send heat through space ... only radiation, and that radiation causes any mass that it encounters to heat up. Not easy to get your head around, I know; it's confusing.

Starting from scratch:

1. Ill-understood processes going on inside our sun cause the outer atmosphere of the sun to be at nearly 6000K (some 5700°C)for full details: http://www.universetoday.com/18092/temperature-of-the-sun/ Keep in mind that what really goes on at the heart of our sun is by no means a settled matter! What is settled is that the outer atmosphere is at around 6000K.

2. All matter radiates according to its temperature. The sun radiates according to its 6000K outer atmosphere.

3. The frequency and strength of this radiation determines the amount of heating it can cause when it hits matter.

4. The amount of heating depends as well upon the distance from the source. You'll understand that if you were only one kilometre from the sun you'd be close to that same 6000K; and on the other side of our solar system, Pluto gets only a fraction of the heat.

Now to our atmosphere.

1. Sun heats earth

2. Earth heats atmosphere

3. Hot air rises, mostly by convection, as you say already

4. There comes a point where the air get so thin (the molecules that make up air are getting more and more spread-out, which is the reason for adiabatic cooling) that convection stops and the only way in which energy can still be lost is radiation.

5. The level of that radiation is in line with the temperature of the emitting gases, in earth's case some 240-Watts per sqm.

Remember that the atmosphere is heated from the bottom up yet it can also be said that the rare atmosphere above the troposphere is heated by the sunlight and reaches high temperatures! But ... there are so few molecules of matter at those high altitudes that no heat is perceivable; it is already close to an absolute vacuum up there!

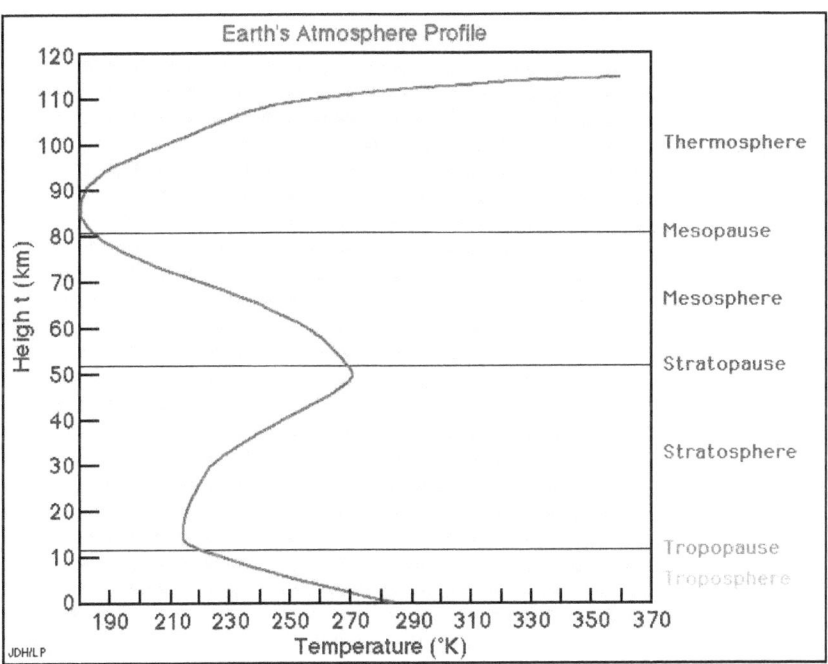

The apparent heating up in the stratosphere is caused by clouds forming and a gigantic release of energy from water vapour turning into water (540 calories per gram of water; you only need 100 of those same calories to bring water up from zero C to 100C, so you can know how much energy 540 calories is).

The heating in the thermosphere is deceptive as it is a near-vacuum up there and only the odd bit of matter reaches those temperatures. Balloonists going at great height always need protection from the cold!

Heat has no temperature; heat is the result of an action. Radiation also has no temperature either; it is the result of a substance having a temperature. Very confusing, I know!

Think of a hotplate on a stove. Before you switch the power on, all is at equal temperature with the rest of the surroundings. Now you switch the power on. Electricity (which has no temperature either!) starts to flow through the element and *causes* it to warm up, which *causes* it to shed radiation according to the temperature it is achieving due to the electricity flowing through it!

Yet, the electricity has no temperature!

Hope this helps; am going offline for a few hours now.

Hans

If they were serious

Anthony Bright-Paul
February 19th 2012

If they were serious, if the followers of Al Gore and James Hansen and Michael Mann et al; if the Hollywood crowd and those who sincerely wished to Save the Planet, if those who assert that emissions of Carbon Dioxide are going to cause such Global Warming that the Arctic would disappear, Manhattan would be flooded and Disneyland would be inundated – if they were really serious, why do they not start nearer home, in California, in Yellowstone National Park?

Why start with the emissions, the piffling amounts of CO2 that man produces, when on their doorstep they have these giant geysers, which blow a mind-boggling amount of Carbon Dioxide into the atmosphere with incredible regularity almost on the hour?

> The Mammoth Hot Spring at Yellowstone, USA, pumps out 160 to 190 tonnes per day of CO2.

Per day! I knew that I had read this somewhere; I believe I have even written something like this before, but nowhere could I find these figures, not in my own writings, not in any searching in Google.

At last I hit upon it – that massive tome of information "Heaven+Earth" by Professor Ian Plimer. If you are even half a Skeptic, that is to say if you have some doubts, if you need facts and figures, then read this book, again and again and again. On p.217 you will find the above quotation, with notes as to the source.

Here is another morsel, under Volcanic gases: - p.216

> White Island, in the offshore extensions of the Taupo Volcanic Zone of New Zealand, has many active and fossil geothermal systems. Every day it pumps out 4,800 to 18,000

tonnes of H2O vapour, **1150 to 4120 tonnes of CO2**, 320 to 1200 tonnes of Sulphuric acid...

Once anyone begins to realise the enormous quantities of Carbon Dioxide that is produced by geysers, fumaroles, vents and so on **every day**, then and only then can one realise the puny amounts that man is adding to the atmosphere.

Clearly the Holier-than-thou crowd could make a massive first step by attempting to bung up all the vents and geysers in Yellowstone! Ha! Ha! I guess I must have a malicious streak. And we have not yet come to volcanoes. Why! I mentioned that a volcano could produce in one burp enough CO2 to eclipse the total of man-made CO2 for an entire year. A young lady wrote to me: -

Your point about the volcano seems like a clincher, Tony!
I had no idea...

Love, Monica

Thanks for the love, Monica; I am but a humble scribe. Say thanks in your heart to Professor Plimer, whom I am quoting, since my expertise is entirely second- hand. But is not that short note revealing? How many people must be taken in by the propaganda from the Greens, by totally idiotic and unfounded assertions that the Global temperature will rise by x number of degrees, leading to all sorts of horrors?

Some of them even believe their own propaganda. Yet more than 80% of the world's volcanoes are submarine – that is under the sea. Our Met Office cannot see any other reason for the Warming other than man-made Carbon Dioxide. Well, well that is funny, as everyone agrees that atmospheric CO2 has increased, but even the CRU has had to admit that no warming even by their arcane standards has occurred for a dozen years. But not one of these famous scientists are able to monitor exactly how much Carbon Dioxide emanates from these volcanoes and the numerous oceanic vents.

Only the lunatic Greens claim an equally lunatic consensus, poisoning the body politic, so that Princes, Prime Ministers, Chancellors, Bishops and Hollywood Film Stars all partake of their scientific claptrap.

Every day as the sun shines on the oceans Carbon Dioxide rises in a stream into the atmosphere, and every day when it is colder the Carbon Dioxide sinks into the oceans, a huge and never-ending exchange. Every single day the massive quantities of phytoplankton by an exercise in photosynthesis turn Carbon Dioxide into Oxygen for the fishes, the fowls of the air and ultimately for us human beings. The whole Carbon Cycle is a wondrous thing, a miracle of Nature. The checks and balances of the Natural World are even today but dimly understood. And the wisest and humblest of our scientists acknowledge just that.

Every time I read that claptrap about Carbon Footprints, and how to reduce them, my gorge rises. How can these ignoramuses have the gall to inculcate innocent schoolchildren with this hideous bile? For that is what is happening. Science is being distorted, textbooks and even Wikipedia is being suborned, so I believe, in order to fit in with a maniac view of the universe. Some of these cranks would even seek to sequester Carbon Dioxide underground, which would be really funny, - in view of the fact that Carbon Dioxide in the aforesaid vents and geysers is expressly attempting to escape from the Earth's interior, - were it not so pathetic and so tragic.

One thing becomes clear as we observe the figures for the atmospheric gases. Nothing is fixed. All the figures we have are approximations, are sometimes useful averages. But that is all. The atmosphere is in a constant state of flux, and that is the only constant that there is about it. Every second of every day in every place change is occurring. This Earth is hurtling round the Sun on an elliptical course. Sometimes we are farther away from the Sun than the distance of the Earth to the Moon – spinning like a top and wobbling on the way. Does this make a difference? Surprise! Surprise! When we are closer to the Sun we get hotter.

What else? Let us not forget the winds and the waves. Sometimes it blows a gale, sometimes the seas are calm like a millpond, and sometimes the Moon lifts the waves in tumultuous storms. This is all so obvious that it would seem to be unnecessary to have to declare that we live in a dramatic universe. Nothing whatsoever is static. Nowadays with time-lapse photography we can observe the clouds forming and re-forming, we can watch the stars racing across the Heavens. And if we could have arranged a time lapse photograph of our own individual lives we would be forced to observe the progress from birth to maturity and from maturity to old age and then death itself.

Then we get rain. Rain is a cosmic miracle. Not only could we not live without drinking water, but rain also washes the atmosphere and clears it of dust particles, aerosols and unwelcome gases. Yet we have these pathetic bigots who call themselves environmentalists whose only real interest in the environment is to make political capital.

If these bigots really believed that Carbon Dioxide was the cause of Global Warming and Climate Change, if they were truly serious, would they not by now have blanketed the oceans, waged war on bacteria, stifled the geysers, and forbad us humans to breathe out! Is that ridiculous? Yes, indeed that is how ridiculous it is.

Anthony Bright-Paul
February 19th 2012

Heat Generation Revised

Anthony Bright-Paul
Tuesday, 10 April 2012

The most important point about heat is that it has to be generated. In one form or another heat is created from combustion. The clearest and most obvious example is the Sun. The second most homely example is a fire, a bonfire. Even with wind turbines, even with hydro-electrics, electrical energy has to be created. Whether heat is created from friction, or whether from the more arcane elements of fusion and fission, whether by man or by Nature, heat has to be generated.

So while heat has to be generated, on the other hand most cooling occurs naturally. Everything is being heated up, or it is simply cooling down, dispersing its heat to anything and everything that is less warm. Of course, man can accelerate cooling with air-conditioning and refrigeration. However, most cooling flows naturally.

The Earth is warmed by the Sun's radiation. This is so obvious that it seems unnecessary to state it. But it has to be stated because of the present delusions with so-called Greenhouse gases.

A gas is a substance. Coal is a substance. Oil is a substance, or a commodity. In order to produce heat these substances have to be ignited and burned. None of these substances produce heat without combustion.

Temperature is a measure of hotness; that is heat. There is a misconception that heat can be trapped, and that is because in a confined space that is heated, temperature will rise. So in the example of the Warmists, the example of the sun shining on a parked car with its windows closed, the temperature within will rise. But equally the temperature will fall once a dark cloud passes, or night falls. Temperature will rise in a confined space **only as long as heat is being generated.**

Temperature can be increased to high degrees in furnaces. A smith can burn anthracite and with the help of bellows pumping in oxygen a temperature is achieved sufficient to make iron red hot, so that a horse's shoe can be fashioned. But we all know that once the shoe is out of the fire it cools, and the fire likewise. A horse could hardly run with a red-hot horseshoe. The fact that some substances and some gases cool slowly does not mean that heat is trapped, though most gases I understand release their heat in nano-seconds.

My Welsh Sceptic friends get very excited by the evidence from the CRU that there has been no Global Warming for over a decade. But, my good friends, there has never been any **man-made** Global Warming ever. Never. Let me repeat that. There has never, ever, been man-made Global Warming and there never can be such an item. It is utterly impossible for man to heat the Globe in any form whatsoever.

The only thing that can warm the Globe, our planet Earth, is the Sun. All the variations that we enjoy, whether it is too hot or too cold, whether it is windy or whether it is calm, whether it is dark or whether it is bright, everything depends upon the Sun. To talk of Anthropogenic Global Warming is the biggest misnomer ever. **Only a star, only something with the magnitude and the power of our Sun, could possibly warm the Globe.**

The Earth is orbiting the Sun on a huge ellipse. Nobody disputes this now. The Earth is tilted at an angle, which accounts for the seasons. The Sun at its perihelion, which is at its nearest point to the Earth, is calculated at some 91 million miles. At its aphelion it is calculated at 95 million miles. Thus at times we are some 4 million miles further from the Sun and at others some 4 million miles nearer. Those facts alone can account for any warming and cooling. The Sun is the most mighty generator, with a surface calculated at 6000K or some 5600C. Its radiation traverses the vacuum of Outer Space. Is Outer Space then mighty hot? Answer: No, because there is nothing there to get heated. On the other hand any matter, any mass that strays there will get heated according to the proximity.

The Sun's rays pass through the four levels of the Earth's atmosphere. Both Alarmists and Sceptics agree to this. That means that the Nitrogen, N2, and the Oxygen, O2, are transparent to this radiation. They do not get heated. On the other hand the Earth and the Oceans are warmed by the radiation, as radiation encounters mass. And then the surface of the earth and the oceans warm the atmosphere.

What is obvious to a Physicist is not obvious at all to the man in the street, the layman. It is natural for the layman to feel that the Sun heats the atmosphere, and the atmosphere heats the Earth. Most of the supporters of the Warmist scientists, I suspect, have this misconception, a misconception that allows them later on to fall for the theories so pervasive at this present time of Greenhouse Effects. The idea that gases in the atmosphere could have the slightest effect on ground level temperatures could only arise for the layman if he already held the misconception that the atmosphere heats the earth, and not the other way round.

Because of the atomic structure of Nitrogen and Oxygen they are transparent to the Sun's radiation. That is agreed by all parties. The incoming radiation however can heat molecules, that is, where two atoms are bonded together. We know this from the Thermosphere, where the heat content may be minus 80°C yet the very few molecules can reach 1400°C. Near the surface of the Earth the air is dense. So the main heat exchange mechanism is conduction as the atmosphere wraps itself round every surface of land and sea.

So heat exudes from the Earth's surface and rises and as it rises it cools. That is self-evident. Well, if not self-evident, it is evident to anyone who has observed that heat always rises in a liquid or a gas. As the molecules rise they lose their heat by convection.

The Alarmists, the Warmists, however contend that the Greenhouse Gas molecules trap the heat. Where? At ground level, where they are still warm? But we know from a Stephenson Screen, which may be 5 foot off the ground, that the readings will already be cooler at 10 feet off the same ground. Can something cooler re-radiate back on to its source; can it radiate back to something warmer? That defies all logic. It is totally bizarre.

It is sometimes argued, as in the Met Office website, that the Earth would be some 30°C colder were it not for the Greenhouse Gases. That is a strange argument and just goes to show how misleading averages can be.

> Scientists explained the heat-trapping effects of greenhouse gases more than 150 years ago. Research has shown that, without the greenhouse effect, the Earth would be about 30 °C cooler - making it uninhabitable to most forms of life. Because they're so effective in keeping the planet warm, we know that any changes in the amount of greenhouse gases in the atmosphere will affect the Earth's temperature.

Such an argument implies that the atmosphere is static, when it is in constant flux. Even the Standard Atmosphere used by Airline Pilots shows a decrease in temperature of 2°C for every 1,000 feet of altitude. So the Greenhouse Gases that are warmed by infrared likewise cool. They are not keeping the Earth warmer, and the only thing that might theoretically trap heat is a vacuum, such as the vacuum of Outer Space that surrounds the Earth.

If I make tea in a teapot the old fashioned way with boiling water, and then add a woollen tea cosy, will that make the tea hotter? No! It just may inhibit heat loss. If I added a second tea cosy would that make a difference? Would that further inhibit heat loss? Hardly! Greenhouse Gases may inhibit heat loss, like the tea cosy, but they can neither generate nor add heat. Likewise, more Greenhouse Gases, like CO_2, make not a whit of difference to the temperature of the Earth's atmosphere, since it is not an enclosed space. The increasing levels of CO_2 and the failure of CRU temperatures to rise for a decade, demonstrates this point.

We know that the Sun can warm the Earth, for the simple reason that it is hotter. If it were the same temperature nothing whatsoever would happen. The Warmist theory of Greenhouse gases re-radiating back to the Earth from high in the Troposphere defies all logic, let alone the Laws of Physics. Yet I read in one website after another that this absurdity is put forth and even taught to children.

Climate is defined as the average of weather **within a given location**. There is no such thing as world climate. Any simpleton knows that. The weather that makes up the climate of the South Pole is manifestly different to the climate of Montevideo or Buenos Aires. In England the climate of Lancashire is manifestly different to the climate of Yorkshire, by reason of the Pennine hills. The climate of Bengal is completely different to the climate of Tibet. The climate of Brisbane is a far cry from the climate at Alice Springs. Those who talk glibly of *climatechange* are simply guilty of loose thinking. There are a huge number of microclimates on this Earth, which is especially evident in the United Kingdom.

Our relation with the Sun is evidently the prime source of our weathers, which are the sources of our climates. Since the Tropics receive the radiation from the Sun most directly they are the hottest regions on the Globe. Conversely because the angle of incidence is so fine the Poles are the coldest. As the heat near the Equator causes the warmed air to rise, this also causes precipitation in the equatorial regions and also winds, as the hot air seeks to equalise with the cold airs of the Poles. All this is well known. The winds, the trade winds, of which the patterns are well known in turn affect the oceans currents. So the movement of warmed water affects the weather and consequently the climate in various parts of the world.

We all know that weather varies. Once again the Alarmists are at great pains to find a 'human fingerprint' for extremes of weather, when they might look no further than the Sun. The Sun itself is inconstant, with huge solar flares, with mighty solar winds. When this is added to the elliptical swing, the changing tilt of the Earth, the Milankovich wobble, we need look no farther than the Sun and its relation to the Earth for all the extremes of weather and for all the changes of climate that we experience on this planet. In fact we know that the best weather forecasts that we can obtain comes from the site, WeatherAction.com run by the astrophysicist Piers Corbyn, whose prognoses are based on his studies and observations of the Sun and the sunspots.

To sum up, there is a one-way street with heat, which flows from hot to cold. All heat has to be generated. Temperatures can be raised in confined spaces - this is not to be confused with trapping heat. Heat cannot be trapped. There is no such thing as a Global temperature – it is a fiction. Mankind could not possibly heat the Globe. Nor can Greenhouse Gases add to temperatures in any way whatsoever. Since temperatures are fluctuating second by second, we are in a state of flux.

The claim by the Alarmists that mankind causes Global Warming or causes *Climatechange* is hereby shown to be fraudulent. *Quod erat demonstrandum.*

Anthony Bright-Paul
Tuesday, 10 April 2012

Comments from the Geophysicist Norm Kalmanovitch

Hi Anthony,

The piece is now absolutely perfect. The only addition that I would like to see is a criticism of the term greenhouse gases because it is a fraudulently fabricated term that has **never been defined to a proper scientific standard.**

None of the named greenhouse gases in the Kyoto accord other than CO_2 and CH_4 (methane) have any detectable effect on the Earth's greenhouse effect because their effect on thermal radiation is outside the band limit radiated by the Earth or have some effect over a low energy portion of the spectrum that is already fully dominated by water vapour.

The effect from methane is so minute that even a ten fold increase in methane would be barely detectable (attached cow flatulence).Only three gases have measurable contributions to the greenhouse effect CO_2 H_2O and O_3 (ozone) but neither H_2O nor O_3 are listed among the named greenhouse gases in the Kyoto Accord because honest facts would be a little too confusing for the public to handle.

What irks me the most about the term greenhouse gases is that most of the named greenhouse gases are in fact pollutants so by lumping in CO2 with these pollutants under the umbrella name of GHG's; greenhouse gases can be named as pollutants allowing the EPA to legislate against 5% of our exhaled breath by referring to CO2 as a greenhouse gas and a pollutant!

It would be great if you could give one more poke at global warming orthodoxy by changing this sentence:

"The Earth is warmed by the Sun's radiation. This is so obvious, so trite, that it seems unnecessary to state it. But it has to be stated because of the present delusions with so-called Greenhouse gases."

To this:

The Earth is warmed by the Sun's radiation. This is so obvious, so trite, that it seems unnecessary to state it. But it has to be stated because of the present delusions with CO2 and the other gases collectively referred to by the fraudulent term " Greenhouse gases."

Best

Norm K.

7 Billion Machines

Anthony Bright-Paul
April 29th 2012

There are presently 7 Billion bipeds walking about on this Earth and they are expected to number 10 Billion by 2050. These 7 Billion machines are you and I. We are factories. We burn Carbon with the help of Oxygen and we produce Carbon Dioxide, night and day.

We bipeds are on the increase. We produce heat. We humans are combustion machines. Without Oxygen we could not burn the Carbon we ingest in the form of food, nor could we produce Carbon Dioxide either.

All heat has to be generated. The Sun generates enormous heat at some 6000K on its surface. A fire generates heat. A central heating system generates heat. Friction – rubbing two sticks together generates heat. An electrical current in contact with a coiled wire resistance produces heat. All animals produce heat, and incidentally produce Carbon Dioxide.

Carbon Dioxide in no way produces heat. It is the **result** of heat production. Both Carbon Dioxide and Water Vapour are coolants, as is well known. They are both used to extinguish fire and restrain combustion.

It is very easy for Sceptics, who talk to each other, to believe that the battle is won, because the arguments against Carbon Dioxide causing Global Warming are so overwhelming. Anyone who reads Climate Depot will see one scientist after another coming to the Sceptic side. **'James Lovelock fallout – Warmists into their last days'**. The news from Benny Peiser on 'fracking' is encouraging on the availability of fresh sources of natural gas. The arguments employed by Johnny Ball against

Wind Turbines are overwhelming. It would seem that the battle is almost over, and victory for common sense is in sight.

That is partly because, for most Sceptics, reading The Guardian is nauseous. Alas we have to read such things as the Guardian and the Met Office website to know the arguments of the Alarmists. They make dispiriting reading. Here is a headline from today's Guardian.

We can have safe, sustainable energy

With renewables we can contain consumption – and climate warming at 2C – if the big users act now

Maria van der Hoeven guardian.co.uk, Tuesday 24 April 2012 22.50 BST

In spite of all the evidence from the Climatic Research Unit itself that no warming has occurred for a decade, the warmists stick to the idea that they must contain warming at 2C. How can we deal with people who live in cloud cuckoo land? And just who is this Maria van der Hoeven?

The arrogance of human machines imagining that they could contain the temperature of the Earth has to be seen to be believed. In the first place there is no World temperature. There is no huge thermometer which one could stick under the tongue of Mother Earth, no armpit to take a reading. Oh! I hear, but scientists are taking readings from thermometers all over the place and they are working out averages.

If the average temperature at the North Pole is minus 30C and at Accra it is plus 30C, what is the average? The average is meaningless. There are thousands and thousands of micro-climates. Everywhere we live on this Planet the temperatures of our given locations are varying minute by minute, second by second, according to sunrise and sunset, according to the seasons, according to the angle to the Sun and its distance from our Planet.

We live on a huge sphere. At any one time half the sphere is in darkness, while the other half is in light. But that again is only half true, because the sphere is revolving. Nothing is static. So it is warming and cooling. Futhermore only one part at any one time is directly perpendicular to the Sun's radiation. The tropics receive much direct radiation and the Poles hardly any at all.

So some say, but look at the glaciers. Yes? Some glaciers have retreated over the past 30 years, but equally some have advanced. But so what? If some warming has occurred in some regions, what is the cause? The Alarmist leap on Carbon Dioxide, or they lump together Greenhouse Gases. But, with what justification? It is not a scientific response, it is simply hysteria.

While we are about to enter the coldest May for many years according to the astrophysicist Piers Corbyn, the Alarmists point to the North Pole and the Arctic ice. But the latest reports show that Arctic Ice cover is at maximum extent. It ebbs and flows with the seasons. Nothing in Nature is constant.

The assumption that any supposed warming is caused by Carbon Dioxide is simply fraudulent. While it is true that Carbon Dioxide absorbs infrared, it is also true that it cools as it rises with the adiabatic lapse rate. So it cools by conduction and convection. So much is evident and well known to any layman. For that reason the Alarmists have to invent another totally spurious idea called 'back radiation'. In fact it is on this back radiation that the whole theory of man-made global warming rests. But when this idea is examined critically it simply collapses. As hot air rises it cools as the molecules grow farther and farther apart. So the extra warming, which is supposed to be caused by greenhouse gases and which is in fact not happening, is purported to arise from back radiation, - back convection and back conduction being naturally ruled out.

This back radiation presupposes that a body can heat itself. Tell that to a pensioner dying of the cold. Heat, so they say is trapped. That is supremely false. Nowhere, no way, at any time anywhere is heat trapped. It only needs a little thought and a little common sense and common observation to realise that this scenario is completely false.

Tell me one single place that heat is trapped. Heat has to be generated. Heat is always flowing from hot to cold. Heat is only maintained while heat is being constantly generated. Watch a bonfire. As the smoke rises up, so does the carbon dioxide. Within barely 100 feet it has lost its heat. Are these cold and rapidly cooling molecules supposed to re-radiate a heat that they have not even got? Heaven preserve us from such chicanery!

In every case it is clear that it is the activity of the Sun or its lack of activity that has the most profound effect upon the climates (note plural) and the weather systems of the world. It is clear that a quiescent Sun produced the Little Ice Age that lasted some 450 years until approximately 1850. A more active Sun has produced the present warming, such as it is, and the campaign against Carbon Dioxide as a cause is simply spurious and non-scientific. Only something of the magnitude of the Sun could possibly cause Global Warming.

The Met Office argues that the Greenhouse Gases trap heat. Effectively they are saying that the Greenhouse Gases act as insulators, meaning that they delay the exit of heat. That may well be, but heat is not trapped. If an insulator delays the exit of heat, far more important is the observable fact that they obstruct the entrance of heat from the Sun. We can see that this is an observable fact by watching the effect of double-glazing. Double-glazing on a conservatory reduces incoming heat far more that delaying the exit of heat. It is the same principle for the atmosphere. The atmosphere reduces the effect of the Sun's radiation far more than it impedes the exit of heat from the Earth. The argument that Greenhouse Gases cool the atmosphere is proven by simple observation. And the huge swings in the temperatures of the Moon prove that the Earth's atmosphere does more to keep us cool than it does to keep us warm.

The Warmists, the Alarmists, in spite of everything, have still won the public debate, by the simple repetition of lies, reinforced by Government funding. The Skeptics have won the scientific argument against anyone who is prepared to enter the lists. But the Alarmists will not argue the case for fear that they will certainly lose.

The man-in-the street will never be convinced by scientific arguments, which are above their heads. What does 'radiative forcing' mean to the average man? Or the argument about whether the atmosphere is a 'black body' or not? How many people have even heard of the 'adiabatic lapse rate'? In the face of such terminology the common man rolls over and leaves it all to the 'scientists'.

Unfortunately the scientists who have had the best propaganda machine are the Alarmists. The idea that we must save the Planet, which will be doomed if we were do nothing, was a mantra that was easy to sell, especially as there was a Government remit to establish by hook or by crook Anthropogenic Global Warming. When this was supported by the BBC, the Met Office, the Climatic Research Unit and a howling mob of Gore-sprung activists, the odds were stacked against the Sceptics.

Since everyone feared for their jobs that did not toe the Party Line, it is small wonder that the 31,000 Skeptic scientists are largely made up of retired Professors. As Tim Ball says in his 'Analysis of Climate Alarm' in the opening Chapter to "Slaying the Sky Dragon": "The challenge facing anyone trying to counter the exploiters is to bring logic, clarity and understanding in a way that the majority of people can understand."

It is for that reason that I attempt to bring in homely references, which are easily understood. If a house is insulated with loft insulation and the cavity walls are filled, will the house be warmer or cooler? A well-insulated house will actually be colder vis-à- vis the sun's radiation, but it will be warmer inside <u>if and only if some heating is generated.</u>

This is a homely illustration, which in fact mirrors what the atmosphere does. But it must be remembered that the radiation from the Sun is infinitely greater than the infrared radiation coming from Earth. In this way it is clear that the atmosphere, in which Water Vapour and Carbon Dioxide play a part is to preserve us from the extremes of heat and likewise to delay the exit of heat from the Earth. In this way we can see that the Greenhouse Gases, far from having a warming effect, in truth have a net cooling effect.

Everything by itself is cooling; all heat is flowing from hot to cold. This is a fact that everyman can observe, whether because the radiators in a house cool once the boiler switches off, or because red hot lava cools to black basalt. There is nothing whatsoever in Nature that grows hotter and hotter, unless heat is being generated to make it hotter.

Since I declare to all and sundry that I am a Climate Sceptic, I engage in all sorts of disputes. Two things stand out. But, Tony, what about the Arctic? But, Tony, I have seen glaciers, as they were thirty years ago and how they are today.

Sure, there has been warming, which we can see in hindsight. From about 1650, which was the depth of the Little Ice Age, the Earth has been gradually warming. That much nobody will dispute. It is clear that the freezing conditions of the Little Ice Age and the subsequent warming had nothing whatsoever to do with man, but with the forces of Great Nature. It is abundantly clear that the cooling and the warming were directly related to the activity of the Sun.

As to the Arctic Ice this varies from season to season. This variation has much more to do with the flow of warm ocean currents than with Carbon Dioxide in the atmosphere. As to glaciers, this is a study in itself. There may well be some glaciers that have retreated, but the dramatic pictures of glaciers calving, with huge chunks falling into the seas, is evidence of advancing glaciers, not retreating.

All these phenomena of Great Nature can in each and every case be assigned to natural causes, and the biggest cause is the Sun, supplemented by geothermal heat and the warmth from the oceans.

This is something that is agreed by all independent scientists. Ah, you might say that that is a colossal sweeping statement. Yes, it is true, because I distinguish between propagandists and scientists.

All those whose remit is to find a human fingerprint are propagandists. They are not scientists in the true sense. When people say, 'Scientists say …' and you can see these words on the website of the Met office, what is really meant is that 'Propagandists say…'

Propagandists in the form of scientific advisers have the ear of many Heads of State throughout the world, with the result that there is creeping mange over the Earth of useless wind turbines and solar panels. Where it is sensible to look for the cheapest and most available forms of energy creation, governments have been persuaded to look for so- called sustainable solutions. And the only thing that is sustained is mendacity.

Anthony Bright-Paul
April 29th 2012

History of Wrong Conclusions

Anthony Bright-Paul
Tuesday, 26 July 2011

The History of Science is the history of wrong conclusions, being eventually overturned by more evidence, or more correctly by logic. Even today, the data that science uncovers is more often than not misunderstood. We have only got to go back to Copernicus and Galileo, or more recently to Milankovich, to see how the scientific establishment in each era resists the correct conclusions, and this same resistance is apparent today. I hate to say that Religion nearly always opposes Science and scientific truths, so I will amend that and say, without fear of contradiction, that False Religion almost always opposes Philosophy, that is the search for truth.

That False Religion opposes Truth is no more apparent in the past than it is today. In fact what we are truly concerned with is Philosophy and not Science. Science is the study and establishment of Facts; Philosophy is that which enables man to come to correct conclusions. Today, with a welter of scientific facts, a truly terrifying mountain of wrong conclusions surrounds us.

I will attempt here to show that every layman has every right to examine the data and form his or her own conclusions. There are those who say, 'We must leave it to the experts'. Or who say, 'I don't know enough about the subject'. But need we be supine? We have been endowed with Reason, and should use it!

So in the foregoing I will attempt to stimulate every reader, not to come to my conclusions, but to come to their own, and I will do this simply by a series of questions. I may put in the answers, where they are incontrovertible, or I may do that later. I want to ask you a whole series of questions about Climate and Climate Change, since there seems to be so much confusion and misunderstanding about this subject.

Does man control the Sun?
Does man control the sunspots on the Sun?
Does man control the solar winds?
Does man control the solar irradiation that reaches the Earth?

Those are my first questions. Have any of you answered 'Yes'?

Does man control the winds?
Does man control the speed of the winds?
Can man control the Jetstream?
Can man control the barometric pressures?

Anyone here for 'Yes'?

Does man produce clouds?
Can man produce rain?
Does man control snowfall?
Does man control hailstones?
Can man control monsoons?
Can man produce aridity?

I don't know about you, but the Met Office doesn't seem to be able to control anything!

Does man control Outer Space?
Does man control the Stratosphere?
Can man control the Troposphere?
Can man control the rate of heat loss in the atmosphere?
Can man alter the Adiabatic Lapse Rate?

There are some who might answer 'Yes' here. Come to your own conclusion.

Can man control the seas and the oceans?
Con men control the height of the waves?
Can man control the Tides?
Can man control the salinity of the oceans?
Can man control the depths of the waters?

Does man control the Gulf Stream, La Niña and El Niño?

Or the Pacific Decadal Oscillation?

Anyone go along with King Canute?

Does man or can man control Volcanoes?
Does man control Volcanic eruptions?
Does man produce Hot Water Vents and Geysers?
Does man control ocean floors and subsidence?
Does man control the movement of the tectonic plates?

Anyone for God?

Can Man control the elliptical path of the Earth round the Sun?
Can man control the angle of the Earth to the Sun?
Can man control the distance of the Earth to the Sun?
Can man control the Milankovich wobble?
Can Man control the Earth's rotation?

Can Man produce lightning, electrical storms and tornadoes?
Can Man produce hurricanes and typhoons at will?

I think that to all the above questions the answer has to be a resounding, 'No'. Man cannot and does not control Climate in any way, shape or form. However, some may maintain that Man does influence Climate in various ways, so we will try to examine that in this same way by asking a series of questions. I may have to put these questions into some sort of context to make them sensible. And the answers may be Yes or No, depending.

These questions concern the properties of gases. The question here is whether gases are active or passive?

Can Carbon Dioxide be frozen?
Can Carbon Dioxide be liquefied?
Can Carbon Dioxide be cooled?
Can Carbon Dioxide be warmed?

In this case I will suggest that the answer in every case must be, Yes. Carbon Dioxide can be made into Dry Ice, which is even colder than Water Ice, and can even cause frostbite. Carbon Dioxide can be liquefied and is often so done for ease of transportation. Carbon Dioxide can be cooled, as in Ice-Cold lager from a fridge. Carbon Dioxide can be warmed as in warm beer. So the question is this: Is Carbon Dioxide active or passive? Please note above the use of the passive tense.

Let us do the same with Water Vapour?

Can Water Vapour be cooled or even frozen?
Can Water Vapour be warmed?
Can Water Vapour retain heat for some while?
Will Water Vapour cool by itself?

Again here, the answers must be Yes in every case. Freezing fog is Water Vapour and maintains cold. Humidity is Water Vapour and retains heat. It does not produce cold, nor does it produce heat. So the answer is that Gases are passive.

Can gases be compressed?
Will gases expand to fill a container?
Can gases be inhaled and exhaled?

I hope so - in all cases it is clear that gases are passive. They re-act. In no way can a Gas jump out of a Gasholder or a can, like a Genie, and say 'Tickety-Boo!' (I am willing to be corrected!)

The next questions are about the Environment.

Does man affect his environment?

The answer to this must be simply, Yes. Man is born without clothes. So man immediately affects his environment by weaving cloth and by using animal skins. So man affects his immediate environment. Man also affects his environment by cutting down trees and clearing ground and growing crops. He also plants trees. Man bakes bricks, makes cement and builds houses, factories and temples. Man diverts rivers and builds dams and canals. Man builds sewage systems. Man uses pesticides to clear away mosquitoes and other pests. In other words in a million and one ways Man adapts to what is most often a hostile environment. Man burns wood, coal and oil to produce heat.

Man can heat water for a variety of processes including cooking of food, and central heating in buildings.

Man also creates laws in order to protect the environment, to protect rivers and seashores from a variety of pollutants.

In fact, in order to live and to be mobile, Man, from the beginning of time has had to adapt to his environment. He has smelted iron, formed bronze from copper, has used lead in a variety of ways, including roofing. He has built roads, and dykes. He has made leather from animal skins. He has milked cows and goats and made a variety of dairy products. He has made medicines and hospitals for the sick. He has also made explosives.

So here it is abundantly clear that man has affected the Environment. That is abundantly true and is clearly within Man's remit.

Some people will argue from this that since Man affects his environment, that he therefore in some ways alters the Climate. Is such a conclusion valid? I am afraid it looks tempting so to conclude, but it is a ***non sequitur.***

Prove it to yourself.

Man does not drive climate, but he adapts to the drivers of climate.

The Mid -West of the USA is an alley for Tornadoes. Some tornadoes are over a mile wide and maybe 50 miles long. . It is interesting to look up in Google the cause of tornadoes, so the question is:

Does man create Tornadoes?
Are Hurricanes man-made?
Are Thunderstorms and lightning man-made?

I am afraid that in these days of super self-delusion there may still be some AGW cranks, who will aver that Hurricane Katrina was man-made, or caused by the combustion of fossil fuels. And they will somehow twist the facts and aver that almost every extreme climatic condition is somehow the fault of wicked man. But it is absolutely clear to any normal, logical person, that all these extreme features of Climate and of Weather simply occur, naturally, beyond the power of man.

Where tornadoes are likely, it is perfectly feasible and desirable that at least half the house is built under ground level, that is, with a huge basement. In the same way where there is a known flood plain, then houses should be built on stilts, as is already done in some deltaic regions. In fact Local Authorities could and should demand that all houses in flood prone areas are built above garages, and the services of gas and electricity should be above the reach of a normal flood, which would also suit Insurance Companies much better.

So though a city like Phoenix may have a colossal urban sprawl, such that it take hours to drive out of it, as I know full well, it can hardly be argued that the citizens of Phoenix are affecting the climate. On the contrary, such is the heat there that most every house and Hotel has air-conditioning, which is a way that man has adapted to the local climatic conditions.

To suggest that the adaptation causes the heat is to confuse cause and effect. That is why the present arguments about Global Warming and Climate Change are essentially not scientific arguments at all.

Furthermore we know that the 'Standard Atmosphere' used by airline pilots throughout the world, takes the average surface temperature as being 15°C, with the rate of lapse, that is the decline of temperature by altitude as being 2°C for every 1,000 feet. That means that at 7,500 feet the temperature is zero, 0° centigrade. Any travellers on a modern aircraft can confirm this for themselves, as the monitor will show decreasing temperatures, so that at 30,000 feet the temperature is circa Minus 45 Degrees Centigrade.

What conclusion, what logical conclusions must follow from that? If gases are passive, if gases can be warmed or can be cooled, if gases have no inherent temperature of their own, then there is no way that they can cause warming. They are either warmed or cooled.

If we put a potato in a microwave oven and switch on the power, the potato can be baked within 10 minutes. But put the same potato in a freezer, how long will it take to bake? Is there a hot spot in the freezer? And yet our noble scientists have been searching for a Hot Spot at 10 Kilometres high in the Troposphere, that is some 33,000 feet! It is not a question of Science it is a question for Logic; it is a question of Philosophy.

The crux of the argument is philosophical and the nub of it is logic. To suggest that Man is somehow creating radical changes in Climate is an inadmissible conclusion. And as to Man warming the Globe, it is a complete impossibility.

Anthony Bright-Paul
Tuesday, 26 July 2011

Living on the Moon

Anthony Bright-Paul
Tuesday, 27 August 2013
(Acknowledging also corrections and additional information from Hans Schreuder and the email from James Peden, Astrophysicist.)

> Temperatures on the moon are very hot in the daytime, about 100 degrees C. At night, the lunar surface gets very cold, as cold as minus 173 degrees C.
>
> This wide variation is because Earth's moon has no atmosphere to hold in heat at night or prevent the surface from getting so hot during the day.
>
> A single "day" on the moon lasts about 28 Earth days, meaning the lunar daytime is nearly two Earth weeks long.
> http://www.space.com/14725-moon-temperature-lunar-days-night.html

Today, August 27th 2013, the hottest place on Earth is Palm Springs, where some dear friends of mine are living – I trust. The temperature has reached 134° Fahrenheit, which corresponds to 56.67°Centigrade. This is reckoned to be an all time record.

Just imagine then living on the Moon, where the daytime temperature can reach 123C and the night time will drop to MINUS 153C. 123C is more than twice as hot as our hottest day recorded.

On the question of temperature I would like to quote an email that I received from the world renowned Atmospheric Physicist, James A.Peden (4.12.2011): -

> "Temperature" is based on a measure of the energy of molecular motion... and indeed, the temperature at the edge of

our atmosphere is quite "hot" ... because the molecules, albeit few in number, have a high kinetic energy ... thus technically have a high "temperature".

However, there are very few of them. Therefore the "heat content" is very small.... resulting in very few calories per unit volume. At sea level, there is a pretty good correlation between temperature and heat content: a kettle of boiling water has both a high temperature and high heat content.

But at the edge of space, with very few molecules per unit volume, you have the seemingly paradoxical condition of both high temperature and low heat content.

Ordinary thermometers work by transfer of heat energy from the surroundings to the thermometer. At the edge of space, they simply don't work because there aren't enough surrounding air molecules to counter the natural cooling of an object by radiation. So, trying to measure the temperature via normal methods results in an erroneously low reading. We must remember that all bodies emit Infrared radiation and thus "cool" in the process. A thermometer may read quite low at very high altitudes not because the surroundings are "cold" but because the thermometer is losing heat by radiation and there aren't enough surrounding "hot" air molecules to counter that cooling.

At the Kármán line ... the so-called "edge of space" (about 100 km) there is in fact an abrupt rise in temperature... as solar radiation reacts with the few molecules still in that region, increasing their thermal energy, and thus raising their "temperature".

Now the Moon like us on Earth is some 93million miles from the Sun, yet it is evidently both much hotter and much colder. Why is this? The answer lies in the sentence 'There is no significant atmosphere on the moon.'

What does this at once tell us? At the edge of space on what is called the Harman line there are very few molecules – like Outer Space the

Thermosphere is almost empty of matter, it is almost a vacuum. But the very few molecules that are there can be extremely hot. (The Astrophysicist James Peden gives no figure for the very good reasons stated in his email.)

The radiation from the Sun has already crossed some 91 - 95 million miles. From the top of the Thermosphere to the level of the sea is a mere 50 or 65 miles, a relatively tiny distance. How then is the surface of the Earth not also similarly hot? Why is the surface of the Earth not 100C –123C like the Moon?

The answer lies in the atmosphere, or rather the fact that we have an atmosphere. Without this atmosphere we should be as hot as the Moon by day and as cold as the Moon by night. In this way we can see that the gases of the atmosphere act as a filter, scattering and absorbing the radiation. In direct sunlight the gases of our atmosphere act as a huge coolant, without which life on earth would be impossible for either man or beast. Out of direct sunlight, as well as at night, those self same gases retain enough heat to prevent the temperature from dropping as low as on the Moon. Even the North and South Poles in the coldest part of their winters do not get as cold as the Moon. Airflows from warmer regions prevent the temperatures from dropping to the same low level as on the Moon.

If we take the lowest level of the atmosphere, which is the Troposphere, where all our weather occurs, we know that the warmest part is right at the surface. At 33,000 feet the temperature is circa Minus 55C. How does all this come about? How is I that we feel warm when sitting or working in the direct sunshine? How is it that cloud cover in the daytime will immediately reduce the effect of the sun?

Radiation has to encounter mass to produce warmth. If we sunbathe we are mass and will experience warmth, even sunburn. The Earth warms up and more importantly the seas and oceans warm. It is the oceans that play the largest part in warming the lower atmosphere, while the atmosphere does very little if anything to warm the oceans.

As the atmosphere warms the warmed air rises, and as it rises it cools. The cold gases cannot heat a warmer earth or oceans, which is the fundamental mistake of the Anthropogenic Global Warmers. The Sun's radiation warms the Earth and the Earth warms the atmosphere, not the other way round.

The Greenhouse Gases play a major role in filtering the radiation of the Sun and thus keeping the Earth cool enough for mankind and the animal kingdom to live on. At night time the Greenhouse gases also inhibit heat loss, so they play a role both in keeping us cool and in keeping us warm, safe from the extreme fluctuations on the Moon. Water vapour in particular acts as a celestial thermostat.

The idea that Carbon Dioxide 'causes' Global Warming is thus seen to be fallacious. The Greenhouse gases do more to cool the Earth than keep it warm, since it is clear that the radiation from the Sun is far mightier than the radiation, infrared, emitted from the earth and oceans.

The idea that Greenhouse gases 'cause' warming is simply not true. Not even the most bigoted Alarmist would claim that these gases generate heat. Nor can they in any way add to the heat that is produced by the radiation for the Sun. Nor can a gas, Carbon Dioxide, trap heat. At most these gases can delay for a very short time the exit of that same said heat, and that would be only at night.

Any layman can learn and understand these principles - from which it is clear that there is no such thing as man- made Global Warming – there never has been and never can be. Far from Greenhouse Gases leading to Global Warming, precisely the opposite is true – without the Greenhouse gases we would all be fried to a cinder!

Shall I tell you a secret?

Anthony Bright-Paul
22.10.2013

Shall I tell you a secret? **The Sun warms the Earth and the Earth warms the atmosphere.** Have you got that? It took me a while to ponder over this. I got it from my Climate guru, my personal trainer in all matters scientific.

If the Sun warmed the atmosphere, the top of the atmosphere, or at least the top of the Troposphere, would be hot – but it isn't – at 33,000 feet it is about minus 55C. I don't have to tell you guys that do I? Just ask anyone who flies and watches the monitor on board. So radiation from the Sun encounters mass and the earth and the oceans warm the atmosphere from the bottom up.

Some smart-arse disagreed with this. He said to me: 'Sit on a cold brick wall on a frosty morning and see just how warm the earth is!' Well that smart arse had a point. So now let me share with you a second secret. **The Sun warms the Earth and Oceans and the Earth and Oceans warm or cool the atmosphere.**

Ah, that makes a difference, that makes sense, doesn't it? When the Sun shines down on the sand even on the Riviera, the sands are often too hot for the feet. Even more so in the Sahara. So the atmosphere likewise gets hot and the temperature rises. When the sun goes down and the sands cool rapidly, even the Bedouin will make a fire and drink hot tea. When the earth cools down so does the atmosphere.

Where it is hot and humid, as in Jakarta, the atmosphere will cool more slowly, but cool it will, inexorably. In the UK now that autumn is upon us and the ground is wet and moist, so is the atmosphere. And as the sun goes down the atmosphere cools rapidly, so it quite cold by dawn.

Is it as simple as that? Of course not! Why? Because we have winds and these winds re-distribute huge volumes of air around the globe.

In the UK the winds play a prominent part in our weather. If the winds come from the south we receive hot air from the Sahara. If the wind is from West crossing the Atlantic we can expect rain and it is often fairly mild. If the wind is from Iceland and the north, yes it will be cold. And when in winter the predominant wind is from the frozen steppes of Russia we can expect a prolonged period of cold and snow.

Shall I tell you another secret that I only learned quite recently from my climate guru? My mentor is a bit like one of those old Zen masters – if I don't get something right he can get quite cross. But then, as I have explained, I just will not believe anything he says unless it concurs with my own evidence.

Just how does the Earth influence the atmosphere? **Here's secret three: By contact, by conduction**.

Let's do an experiment. Boil up a kettle of water, preferably in a nice shiny metal kettle. When it is boiled hover your hands over and round the kettle in order to experience its warmth. Actually you have to hover your hands quite near in order to experience its radiant heat. Now put your hands on the hot kettle! Careful! Now do you see what I mean? If you clasped hold of a hot kettle you would know it instantly. Your immediate reaction would be to let go; otherwise you would burn yourself severely and come out in blisters or worse.

From this extremely simple experiment it is possible to learn two things. By touching, by conduction the transfer of energy is immediate. On the other hand the transfer of energy by radiation is not only slow, but is governed by distance and intensity, which has a fancy name – inverse square law.

What relevance does this have for us in following the climate debate? What relevance does this have for man- made Global warming? It is very simple – we can see easily that the atmosphere warms and cools by touching, by contact, in a word by conduction. And we can also see quite clearly that it cools by convection. As the steam rises out of a kettle spout it is very hot immediately but a couple of feet up it has already cooled. Now the Warmists base all their theory on radiation and here I will append an extract from an email from my learned friend Max Potter: -

> When the energy is re-radiated it is emitted as much lower frequency infrared (below or lower frequency than red). **It is this infrared radiation, which is absorbed by the "greenhouse gases" (including water vapour), which warm the atmosphere.**
> Email from Max Potter Nov 24 2012

Unwittingly Max has explained precisely why the whole man-made Global Warming has not worked and is entirely fraudulent. Firstly it is based entirely on radiation, which as we know is slow and does not travel far. Why? Because we now know from our own observations and experiments that heat energy is moved with immediacy by contact, by conduction. Secondly we know that the sum total of all the Greenhouse gases in the atmosphere is a mere 1%, which includes Water Vapour and Carbon Dioxide as well as other Greenhouse Gases. So this radiation is only being absorbed and emitted by 1% of the atmosphere and arguably only at ground level. Why do I say that? Because of convection, because warmed air rises and cools, and remember the Greenhouse Gases are just a very minor part of the atmosphere. The idea that there should be a hot spot in the sky or a lot of teeny-weeny suspended Chinese lanterns lit up by 'photons' is just inadmissible, if not entirely ludicrous.

It is not just that there is no such thing as man-made Global warming – it is entirely impossible. Heat, by itself, can only flow from hot to cold – 2nd Law of thermodynamics.

Why is it important to establish this fact and to repeat it again and again? Because millions of £s sterling and millions of dollars have been spent and are being spent on an entirely spurious campaign to prevent Global Warming and Climate Change.

Rage, rage against the dying of the light! (Dylan Thomas)

Yes Dylan, I am raging; I am raging against the dying of the light!

Anthony Bright-Paul

Sunday, 20 October 2013

Postscript: Scroll down for 'Contact'.

Contact:

Sometimes it happens that after I have written something the true import of what I myself have written dawns upon me. Of course the three secrets that I have divulged are just the opposite – they are not secrets at all. In fact they are evidential, they are secrets only in the sense that they are so obvious that people fail to recognise them.

Once I had realised that the Earth and the Oceans warm or cool the atmosphere, it was only last night that the enormity of this realisation came upon me. Why so? Because the Earth and the Oceans are in contact with the atmosphere over the whole surface. Or rather if one prefers it, the atmosphere envelops the whole of everything, - every single nook and cranny is covered by the atmosphere.

It does not matter if it is the sands of the Sahara or the vastness of the Atlantic Ocean, it does not matter if it is the Siberian steppes or the mountains of Tibet, it makes not a whit of difference if it is a Silesian salt mine or a New York skyscraper, everything everywhere is covered by the atmosphere. Just everything is in contact.

Just think for one moment what that means. That means that everywhere upon this Earth there is a heat exchange taking place by conduction. So that where there is snow and ice floes the atmosphere cools; where the tropical Sun beats down the atmosphere warms. Where winds blow the air is re- distributed. Only this morning I saw on the BBC News that London had a temperature of 17C last night and why? Because of huge southerly winds bringing up warmth from the tropics.

A lady friend wrote to me over Facebook to say she was consulting a scientist. She has no need to do that. The only need is to consult with the evidence of her own senses. If one goes to the great Cathedral of Chartres or if you visit the Taj Mahal and you were to say that these monuments have an atmosphere, you are correct. It is the same with Pisa and its famed leaning tower; it is the same if you were to visit the Buddhist Temple of Borobodhur in central Java.

If you watch the rushing torrents in North Wales at Betws-y–Cooed, or the placid lakes like Lake Como in Italy; if you see the wonders of the Mer de Glace in the Haute Savoie, you cannot escape the fact that everywhere is enveloped in atmosphere. This enormous covering envelops the whole earth – no place is without it.

You might think that I am stating the obvious. It is so obvious that it is easy to overlook, that every inch of the terrain or the waters of this earth are in contact with the atmosphere, and every inch is exchanging heat with this same atmosphere, night and day, endlessly. Some places are cooling and some are warming, there is no stopping, there is no moment in time that is the same as the moment in time that has just passed.

It is not possible to take an average of this flux for the simple reason that it is constantly on the move. So the atmosphere is warming and cooling and shifting incessantly by contact with the earth. The gases of the atmosphere do not warm the Earth – the Earth and Oceans warm and or cool the atmosphere.

What then of the idea that the Earth radiates infrared at a low frequency and that Carbon Dioxide absorbs and emits it? Sure we will allow that, remembering both the absorption and the almost instantaneous emission. But Carbon Dioxide is a mere 0.04% of the atmosphere. So if it warms up and cools down it is almost irrelevant. My friend Max Potter has conceded that Carbon Dioxide does not generate heat – he opines as many other Warmists do that it traps heat. Really? Tell me anyone, how do you trap heat? In a little black box or a hot water bottle? And would it remain trapped or would the genie escape? Come now, we are not children. There is no way that heat can be trapped.

It can be generated; heat can be generated, even by rubbing two sticks together. So it is suggested that excited molecules rub against each other and pass on their heat, somewhere up in the cold atmosphere and thus cause Global warming. Even if this was less than absurd, is it suggested that 0.04% of the atmosphere has somehow managed to warm the other 99%? The atmosphere cannot warm itself!

Too many Sceptic scientists have managed to waste their time arguing with such tomfoolery, when all the time they could have asserted what is evident, what is obvious to every man and woman of the slightest intelligence, namely that the earth and the oceans warm or cool the planet by contact, by touching, by conduction all at once and all the time, night and day.

Let me repeat this. Every single one of us is surrounded by atmosphere. The land and sea are covered by the air, and the air surrounds us all. Every single surface is either warming or cooling across the entire globe simultaneously.

I will repeat that: - **Every single surface is either warming or cooling simultaneously by contact with the atmosphere night and day**.

Let me ask you, 'Is that credible? Is that observable? Is that evidential?' Any man or woman who is not entirely asleep can observe this warming or cooling by contact every single day. There is no hocus-pocus about watts per square metre; there is no hocus-pocus about infrared radiation affecting molecules suspended in the ether!

No! What I have written and which came to me as a blinding light is easily comprehensible to any garage mechanic, to any plumber and heating engineer and to any housewife who does the cooking.

I am indebted to Hans Schreuder for opening my eyes to such an obvious thing that I have been in a state of excitement with the realisation of it all and the need to pass it on again to all my fellow men and women.

There is no Anthropogenic Global Warming, my friends. It is totally beyond the power of man either to warm or to cool the globe. There is however simultaneous global warming and global cooling and this is entirely in the hands of that great orb in the sky, the Sun.

What's in the space?

Anthony Bright-Paul
Thursday, 31st October 2013

Have you ever wondered when the BBC gives a forecast of temperatures in different regions of the UK, have you ever wondered what those temperatures refer to? With all the argy-bargy about Carbon Dioxide, is the temperature reading - say today at Bracknell between 9C – 13C - does that refer to Carbon Dioxide alone, or does it refer to Water Vapour, or does it refer to Oxygen and Nitrogen, the last two constituting 99% of what we call air?

It seems such a daft question, but since it came to me in the early hours I had to email a friend of mine, who happens to be both a Biblical scholar and a scientist in order to get an answer to such a simple question. Of course, I knew the answer, but I had to be sure that I had not missed something obvious or something arcane that only great scientists wot of. The answer of course is that the temperature is the temperature of all the gases of the atmosphere together, as taken at various stations at some 5ft above the ground.

Since my wife is Manx I always look at the temperature forecasts for the Isle of Man, then for Weymouth, then for the Torbay area, then for Bristol, in those areas where members of my family live.

Both Warmists and Sceptics are agreed that the radiation from the Sun passes through the atmosphere and as radiation encounters mass the earth and the seas warm up. Does that worry you at all? It does me, because of what Galileo averred some many moons ago. He said that the atmosphere has mass. Now my friend Max Potter who has been a consistent Warmist wrote this: -

However, recapping to make it clear, the energy we receive from the Sun is mainly high frequency radiation, not absorbed by the

atmosphere, the energy the Earth emits is low frequency infrared, which is absorbed by the "greenhouse gases".
Email from Max Potter

That strikes me as curious. Why is the high frequency radiation from the Sun not absorbed by the atmosphere or more exactly by the Greenhouse Gases, but the much lower frequency from the Earth is apparently absorbed by these same Greenhouse Gases? Again, why do not the Greenhouse Gases absorb the incoming radiation, which is so much stronger? Or do they? Are these Greenhouse gases picky? It just does not add up, does it? And then again, why does the radiation from the Sun apparently pass through the transparency of Oxygen and Nitrogen, but the much much lower frequency infrared warms these same Nitrogen and Oxygen molecules, according to the weather forecasts? Of course, the answer is that it doesn't. Oxygen and Nitrogen must be transparent to both near-IR and far-IR. So then how does the air get warm?

The Warmists go even further, claiming that the minute quantities of carbon dioxide molecules by some magic warm all the other molecules of the lower atmosphere all at once, which the Sun has been unable to do apparently. There just must be some other explanation.

Now we all know that the atmosphere warms from the bottom up. Why does it not warm from the top down? Yet we do know and everybody agrees on this, that the atmosphere cools by 2C for every 1,000 feet of altitude. So apparently we must accept that the Sun warms the Earth and the Oceans, and they in turn warm the atmosphere. Therefore it must follow that the earth warms or cools the atmosphere from the bottom up by contact, by conduction. And convection follows conduction.

Heat conduction, also called diffusion, is the direct microscopic exchange of kinetic energy of particles through the boundary between two systems. When an object is at a different temperature from another body or its surroundings, heat flows so that the body and the surroundings reach the same temperature, at which point they are in thermal equilibrium. Such spontaneous heat transfer always occurs from a region of high temperature to another region of lower temperature, as described by the second law of thermodynamics.

What happens when convection occurs? The gases rise up and as they rise up the molecules get farther and farther apart and the gas cools. If you are on a dry sandy beach get a friend to throw a handful of sand into the air. Advise him not to look up lest he gets sand in his eyes. As the sand goes up the particles separate in a wide arc. This is what happens to molecules as they rise up. The distance between them increases. We know that climbers on Mt Everest have to take Oxygen with them since there is less Oxygen at high levels than sea level. As the molecules separate so also the atmosphere gets colder and colder.

Have you followed me so far, because now I am coming to the next question? If the molecules separate and are distant from one another, what is **between the molecules? What is in the space?**

When I emailed my friend the Rev Philip Foster he replied within minutes – **Nothing.**

Nothing? What is nothing? Nothing can only be a vacuum, something that is empty.

Here is another take on that: -

At sea level on Earth, we breathe in an atmosphere where each cubic centimetre contains 10,000,000,000,000,000,000 molecules; by comparison the lunar atmosphere has less than 1,000,000 molecules in the same volume. That still sounds like a lot, but it is what we consider to be **a very good vacuum on Earth**. In fact, the density of the atmosphere at the moon's surface is comparable to the density of the outermost fringes of Earth's atmosphere where the International Space Station orbits.

http://www.nasa.gov/mission_pages/LADEE/news/lunar-atmosphere.html#.

Now we know that the radiation from the Sun passes through outer space because it is a vacuum. Outer space has no temperature, only a nominal Kelvin, because there is no mass to get hot. Near the ground the molecules are more closely knit. The earth and the sea are denser, so they heat up (or cool down). The air is warmed primarily by contact, or conduction, which precedes convection.

My good friend replied again with expedition with something that those of you who are scientists might well enjoy. These calculations were done with speed and I was given permission to pass them on with the proviso that they had not been checked.

Right I did an initial sum. Don't quote me yet I may have made a mistake.

Avogrado's number: 1 grm mole of gas has $6.02214129 \times 10^{23}$ molecules at 1 atm. and occupies 22.4 litres.

Diameter of N_2 Molecule 100 pico metre = 10^{-10}m (CO_2 is 116pm)

Molecule Vol $4/3 \times pie \times r^3 = 4 \times 0.5 \times .5 \times .5 \times 10^{-30}$

$5 \times 10^{-31} \times 6 \times 10^{23}$ m^3

30×10^{-8}

3×10^{-7} m^3

Volume of 'gas'
22.4 litres = 22.4×1000 cm^3 conversion 1m^3 = 10^6 cm^3

$22.4 \times 1000 \times 10^{-6} = 2.24 \times 10^{-2}$ m^3

Therefore the volume actual taken up with molecules is $3 \times 10^{-7} / 2.24 \times 10^{-2} \sim= 1.3 \times 10^{-5}$

$= 0.000013 = 0.00013\%$

If we calculate the physical volume occupied by air molecules at 1 atmosphere pressure if they weren't moving about - i.e. if they were piled on the floor of a container like beads, then these molecules physically occupy just 0.00013% of the space, the rest would be 'empty'. Of that only CO_2 and H_2O can actually absorb any IR. As CO_2 is 0.04% of the atmosphere then the space these CO_2 molecules occupy is 0.000000052% of the space.

You did get all that, didn't you? The maths was simple wasn't it? That's good then, because it was way above me. But I'm including it of the sake of those to whom Avogrado is meat and drink.

But the conclusion is what matters. Whether we prefer to think of molecules as billiard balls or beads makes no matter – we are only magnifying them for the sake of convenience. The point is that these somethings that we call molecules are suspended in a vacuum, even though they may be more closely packed at sea level. There is still plenty of space between the molecules. But look at CO_2...**CO_2 molecules occupy is 0.000000052% of the space.**

Since it is not too difficult to observe the energy exchanges that take place through contact, that is through conduction, and since it is likewise simple to observe convection by the simple expedient of lighting a bonfire and watching the smoke and flames rise up, then we can quickly see that the idea that Carbon Dioxide warms the lower atmosphere is a complete non-starter. And where a friend has commented that I did not deal with the Warmists' vaunted back radiation, I did not do so for the simple reason that there is no place for it in the above scenario. Could a frozen molecule effectively radiate earthwards say from 10,000 feet? Bearing in mind the Inverse Square law?

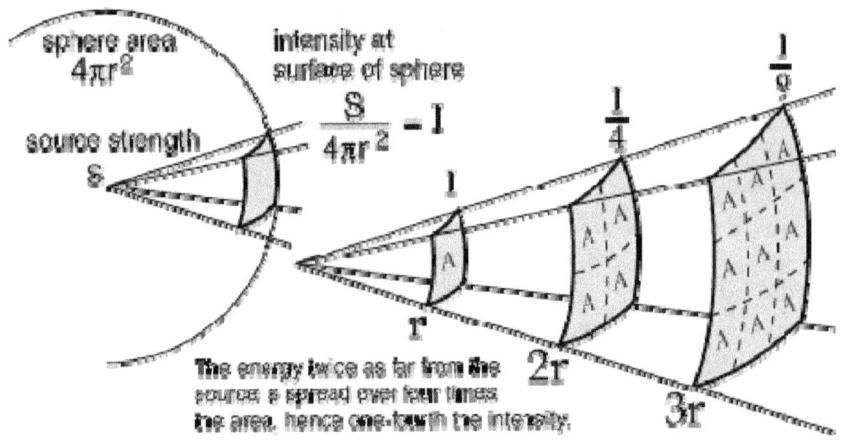

We must remember that none of these gases generate heat. The heat is generated by the Sun. Does CO2 have any part to play? If, like Water Vapour it is opaque to radiation, it must have a 'scattering' effect. That is a net cooling effect, since the radiation from the Sun is some 6 times the intensity compared to radiation from the Earth. Ergo, Anthropogenic Global Warming has no scientific basis.

Anthony Bright-Paul
Thursday, 31st October 2013

You can't heat Nothing!

Anthony Bright-Paul
01 November 2013

Only this morning very early the enormity of what I had written yesterday struck me. Whereas the Rev Philip Foster had supplied some elegant constructs, I will now tempt you with some extremely simple arithmetic. Twice one is two, twice two is four. Are we agreed? But what is twice nought? Of course, it is nought. Let us go further. What is one thousand times nought? Of course, that is also nought, nothing, zero. Let us try one more time – What is 6,000C, which is reckoned to be the temperature of the corona of the Sun, what is 6,000 Centigrade times nothing? – you have it in one. It is zero, it is nix, it is nil, it is absolutely NOTHING!

This accounts for the temperature of Outer Space. Since space contains nothing, then the temperature of Outer Space is zero. Okay, it is given a Kelvin number, since dust particles may stray there. But the space itself is zero. **Correction: Since Outer Space is a vacuum, since it is by definition empty is has no temperature. See the explanation of Hans Schreuder in 'Get your head round this!'**

So now I am going to attempt to answer some of the questions that I posed yesterday in my essay 'What's in the space?' Was Galileo right? I mean, does the atmosphere have mass? The answer is complex, but in the simplest terms the atmosphere has increasing mass from the top down. At the edge of space there are very few molecules and according to an email I had from the astrophysicist James Peden there are very few molecules at the Karman line, but they are very hot – how hot cannot be possibly measured by conventional thermometers. But the vastness of the Thermosphere appears to us as 'cold'. Why? Because it is near vacuum, it is nearly empty. And you cannot heat empty, as per the foregoing conclusions.

So is the atmosphere being heated from the top down or from the bottom up? The answer must be both. The Sun is heating mass, but only where there is mass. So the Greenhouse Gases must absorb the incoming infrared, and this is most easily illustrated by Water Vapour and clouds. For the full scientific argument on how this is done I would refer the reader to the paper by Hans Schreuder, 'Greenhouse gases in the atmosphere cool the Earth.' (Affix link here). Clearly as Hans Schreuder argues, the Greenhouse Gases absorb and scatter incoming radiation, near-IR, which is far greater than far-IR, and therefore the Greenhouse Gases, far from causing warming, do exactly the opposite, - they have a net cooling effect, and this has been confirmed by NASA.

http://www.naturalnews.com/040448_solar_radiation_global_warming_debunked.html

Everything now is perfectly logical. Once we realise that it is utterly impossible to warm nothing, then we also realise that – words are failing me here! - Where there is little mass there can only be little heat, there has to be something to get hot. So the molecules may warm, but the space in between is actually neither hot nor cold – it is absence. We call it cold, because it is the absence of heat. Am I making sense? Shall I tell you something? I do not know myself. I have to ask my scientist friends if this makes sense, as I receive what I understand as a sort of intuition.

The fundamentals remain the same. Radiation has to encounter mass to produce heat. The greatest concentration of atmospheric molecules is at sea level, where earth and oceans meet the air. Therefore what we measure as temperature is but a tiny part of the whole. I leave it to my readers to apprehend the consequences of the foregoing, if proved scientifically correct.

Strictly speaking there is no such thing as cold, but there is absence of warmth, of heat. Which is why I suppose we have the Kelvin scale, which goes down to absolute zero. What we call cold is the absence of warmth because of the absence of matter, the absence of mass, in a word the absence of molecules, since only matter can be warmed.Surely, as the radiation from the Sun descends through our atmospheres it must collide on its way with molecules and divest itself of some of its heating properties, which is why we do not experience the same wild swings of temperature as on the Moon.

Let us be clear then: Neither Man nor the Sun can heat 'nothing'. I ask you then: 'What must this mean for the Globe?

Anthony Bright-Paul
01 November 2013

We also need to remind ourselves that there is actually no physical quantity known as "cold". There is only "heat" and "lack of heat". Astrophysicist Jim Peden

From the 'Great Global Warming Hoax?'

http://www.middlebury.net/op-ed/global-warming-01.html

Note from Tony BP – this is recommended reading!

Can anyone doubt?

Anthony Bright-Paul
10 February 2014

Can anyone doubt that climate is changing? No, because that is what climate does. The weather is changing. As I wake up in the morning I look out at the brook and the level of the water, particularly now as there are flood warnings everywhere. I look at the sky – is it overcast or clear? I watch the news on television, paying particular attention to the weather forecast. I do not need to be persuaded that the weather is changing, because I can see it with my own eyes, and as climate is defined as the average of weather in a particular location, I am also persuaded that my local climate, my microclimate, is also changing.

Is there such a thing as world climate? Well, there certainly is not one world climate – such an item simply does not exist. At this very moment in time, forest fires are raging in southeast Australia, fanned by great winds, which now threatens Melbourne itself. While Melbourne swelters, almost the entire United States of America and Canada is snowbound. The severity of the weather is on every news channel, such as CNN and Fox. In the United Kingdom we have had now 6 weeks of storms and the most severe flooding that most of us have ever known. Europe is likewise afflicted. Indonesia at the other end of the earth is also experiencing floods and also volcanic eruptions. So, yes, there are climate changes everywhere – too many to mention – but is there one world climate? On the evidence, there is no such thing.

What is the cause of climates changing? It seems to be agreed that the Jetstream is blowing right across the world and is more southerly than normal and blowing at great speed. It is this Jetstream in the Stratosphere that brings one low-pressure system after another to batter the British Isles. If this is indeed so, we are faced with a simple question – has man caused the Jetstream to move or to blow stronger, or is this a change that is brought about by Great Nature herself?

Of course the Warmists in the face of natural disasters are hastening to affix blame on mankind. They just hate anything 'natural'. Yesterday on Sky News Dermot Murnaghan spent his time leading a lady Scientific Officer from the Met Office and a Greenie political lady in trying to make Global Warming responsible for our present weather. The Warmists must somehow or other establish that Carbon Dioxide is responsible for man-made Global Warming, and is therefore responsible for heat waves, for winds, for drought, for floods, for ice and hail and snow and of course for sea-level rises.

The truth of the matter is that for the ordinary man in the street it is Mother Nature strutting her stuff, but for the Warmists a natural explanation would be a disaster.

Sure one half of the Globe warms in the daytime and one half cools during the night. That much any sentient being can observe for him or her self. But that is not good enough for the Warmist fanatics. They must have it that man is the cause, for without man to blame, excepting themselves of course, there would be no crusade.

As a salesman for many years I was taught to seek the hidden objection. In our case today we need to seek the hidden agenda. What is the agenda of those who call themselves Green? Whether it is conscious or unconscious the hidden agenda is the destruction of capitalism. The Greens and the LibDems are the Luddites of the modern world. They envy the fruits of capitalism, the wealth and the comforts, the high speed travel, the luxuries of television and IT and all the hi-tech advances, but they have an often unspoken hatred of the capitalist system, which has brought so much wealth to so many, not just in the Western world, but now more and more in the Eastern and the Third World. They may not call themselves Marxists, but that indeed is at the root of the Greens – a driving urge to destroy the economies of the West. They hate big business, they hate Big Oil, they oppose Shale Gas with a sort of religious fervour, but would not hesitate to blame everybody and everything should their gas, water or electricity supplies fail.

While it is evident to any half-educated man is that we are hurtling round the Sun at some 60,000 miles per hour, that our weather and our seasons depend upon the angle we have to the Sun and that this angle varies slightly but enough to cause huge changes on this Planet, these Marxist Greens are intent on blaming Global warming on a gas which humans themselves are contributing to the atmosphere at the rate of 40,000 parts per million for every single breath they breathe. Such is the hypocrisy of the Marxist/Greenies.

They wish to establish that a trace gas is a Driver of Climate. Everything for them depends on this – it is their bedrock. But the truth is other. What is their driver? What is it that drives people to ignore Great Nature and to attempt to blame human kind for just about everything? There must be some reason.

It is not too difficult to pinpoint. The driving force is envy and greed. Look at the politicians who are even now boasting that they have been tackling climate change, while getting themselves sickeningly rich in the process. Need I name names? Politicians of all parties are party to this fraud and some, the most notorious, are in the News right now.

Enough is enough. It is time to acknowledge the power of Great Nature and to do what man has always done – take action to survive. Adaptation is the name of the game and that is what man has always done – adapt.

Anthony Bright-Paul
10 February 2014

The Fundamental Error
Anthony Bright-Paul
Thursday, 20 February 2014

The fundamental error in the entire Warmist thinking is that they imagine that the Atmosphere warms the Earth. As usual they have it arse about face. It is the other way round. Some Warmists search and have long searched for a hotspot in the sky, in the upper atmosphere. Why have they never found it? The answer is simple. Fill a balloon with hot air and what happens? It rises upwards. If there were a hotspot it would fly away, upwards and away.

Sun warms Earth (and I use Earth here to include the waters of the oceans, which cover some 70% of the surface of this Planet), and the Earth warms the Atmosphere. So all the talk of Greenhouse Gases and a Greenhouse Effect is simply 'hot air'.

So my Warmist friends may well be correct as they quote Aarhenuis. One colleague wrote to me as follows: -

In its original form, Arrhenius' greenhouse law reads as follows:
if the quantity of carbonic acid [CO_2] increases in geometric progression, the augmentation of the temperature will increase nearly in arithmetic progression.

That may well be. If the temperature of the gas increases, what happens according to Gas Laws? The heated gas expands and rises. What happens to the surface from which it arises by convection? Why! It cools.

Can any man prove this to himself without taking a degree in Physics? A lot can be done by simple observation, by watching the steam from a kettle or from watching the smoke and flames from a bonfire. The gases always rise up and cool.

Some years ago now I took my wife and daughters to holiday in mid-summer to Megève in the Haute-Savoie, France. Megève is also a ski resort so it is fairly high up. In order to go swimming we would descend by car to Sallanches, the picturesque town on the plain below. There we would swim and sunbathe in the Lac de la Cavettaz, of which one part of the Lac is called the Miroir du Mt Blanc, for the very good reason that while bathing one could watch the summit of Mt Blanc covered in a glorious white cap of snow. How was it that we could be bathing in the warm waters of the Lake, and yet the summit of Mt Blanc was covered in snow?

By all mundane logic the summit, being nearer the Sun, should have been the hotter, but the reverse is true. The reason is simple enough. The air at ground level, or sea level is always warmer than the air at altitude. By what is called Standard Atmosphere we know that the air cools by 2° C for every 1,000 feet of altitude. So, although we were baking in the valley at 86°F or 30°C the air at the summit of Mt Blanc was considerably colder and thinner.

Many bathers sit around and allow the sun to dry them off. Why? As the water changes from a liquid phase to a gas and vaporises, or evaporates, it has a cooling effect on the skin. This is the same principle that applies to all gases on the surface of the Earth. As the gas warms and rises up, the surface from which it rises becomes cooler – not hotter.

Carbon Dioxide far from being a cause of Global warming is precisely the opposite – a huge factor in keeping Planet Earth cool. The warming of the atmosphere leads and must lead to the cooling of the surfaces of the Earth.

Why is it that the air temperature at Torquay on the Devon coast differs from the temperature at Bridgwater, which is presently subject to flooding? It is only just up the road. Why does the temperature at Torquay vary throughout the day and night? Why does the temperature at Torquay vary from Liverpool or Douglas or Norwich, in the east of the country, whose temperatures are also varying minute by minute?

The wind is blowing from the southwest and the south, bringing unusually mild air to the British Isles. The Jet Stream high up in the Stratosphere is also, since December, bringing us one low pressure system after another with massive rainfall and consequent flooding in many areas.

So we have to ask ourselves why do we have winds? Why do we have ocean currents? We cannot surely as a human race be so hysterical in this day and age when we already know about our journey round the Sun to imagine that mankind in some mysterious way has caused the winds to blow and the sea currents to stream?

No, because we know that we are hurtling through space at 67,062 miles per hour on an elliptical course round the Sun, spinning and wobbling on our own axis, an axis that is tilted at some 22 degrees. We know enough to know that we are some 95 million miles away from the Sun but that at the perihelion we may be some 5 million miles nearer the Sun that at the aphelion. In relation to the Sun the Earth is tiny. The Sun is an engine of immense energy and heat, with huge solar flares and mighty solar winds, which impact upon this Planet. To disregard this fact and to imagine that Global Warming and Climate Change is produced by mankind burning fossil fuels takes us back to the days before Copernicus, before Galileo and before Giordano Bruno. Will the warmists next declare that mankind is affecting the Sun? Or the stately progress of the Milky Way?

Why is it that the greater part of the United States is covered in snow, yet Southern California is suffering a drought? Why is it blazing hot in New South Wales, Australia, yet it is blizzard conditions in Japan? Why are there floods in Indonesia and eruptions of volcanoes? If the whole Globe were warming there should be some unity of signals, yet precisely the opposite is the case. Everything, every empirical evidence, points to the fact that there is not one world climate and to talk glibly without consideration of 'man-made climate change' is an idiocy bordering on lunacy.

The reason that it is impossible to talk realistically of manmade Global Warming is because the Atmosphere everywhere over the Earth is in a state of flux. As warm air rises and cold air flows in there is no way that one could affix a temperature to this incessantly moving mass. So that is why Torquay differs from Norwich, which differs from Melbourne, Tokyo and Los Angeles. The temperatures are varying second by second on every single place upon this Earth. There is not one location that is precisely similar to another.

Unfortunately this massive hysteria is costing mankind dearly. The Sceptics have to stand firm and are subject to every sort of vilification and innuendo as they attempt to establish facts and to stand firm by the scientific method. What is really sad is that many well-meaning people, many innocents, have been taken in by the Great Global Warming Swindle.

To talk about tackling or fighting climate change shows a degree of ignorance and arrogance that is incomprehensible in view of the facts and data available. Assuredly Great Nature will have the last word.

Does Wales need a single Wind Turbine?

Anthony Bright-Paul
Sunday, 01 December 2013

My good friend Saxon Aldred, of 62 years he reminds me, writes to upbraid me for my lack of humility. My equally good friend Dr Rachman Michell, also suggests that 'I sometimes feel that in the self confidence and vigour of your reasoning you may just get carried away with the rightness of your reasoning.'

That may well be, but let me ask you a question which perhaps Saxon might answer. Does Wales need a single Wind Turbine?

I understand that from my good Welsh Sceptic friends, who pepper my Inbox with their protestations and copies of letters they have sent to various newspapers, that the gas-fired 2,000 MW Power Station in Pembroke has sufficient capacity not only for the whole of Wales, but also a surplus to be able to export to their old enemy England.

So answer me this. **Does Wales need a single Wind Turbine?**

Please do not protest that you do not have a degree in Economics, or that you do not understand Thermodynamics. If Wales has sufficient capacity to provide for all their needs, why in God's name must their beautiful country be peppered with these unsightly Turbines?

Surely in your travels with your caravan, Saxon, you must have visited Wales. In my days as a Rep I used to visit the Caernarfon area every two or three months. When my youngest was at Swansea University I also got to know the Pembroke coast. How then has it come about that this exquisitely beautiful terrain is being systematically raped? Can you answer me that?

The answer is that it is Government policy. It is Government policy to reward those landowners who will allow a single Turbine to be built on their land something in the order of £ 100,000 per annum for 25 years, plus extras if the wind is too strong and they are forced to shut the wretched thing down. And this policy that emanates from

Westminster is being administered and passed by local authorities in Wales against the manifest wishes of the population of Wales itself. How has it come about that Welshmen can opt against their own countrymen? Does not something smell to you, Saxon? Does not something stick in your nostrils? Is not that sickly sweet scent the stink of corruption?

Perhaps after all I am wrong. Perhaps after all these authorities are only fulfilling the behests of the Department of Energy and Climate Change. Perhaps they are powerless to resist the dictats from Westminster.

So then we must ask, How has this policy come to pass in the first place? Here Saxon, you may indeed need to know one very simple thing, what Rachman calls my kitchen Physics. When you make yourself a nice hot cup of tea from boiling water, does that cuppa get hotter and hotter or does it inevitably and certainly cool? If you get the answer to that right you also know more Physics than Al Gore. Effectively you will have understood the 2nd Law of Thermodynamics without having even lifted a shovel.

Ever since mankind appeared on the Earth man has worshipped the Sun as the giver of light and warmth. Every child who plays with bucket and spade on the beach knows that the Sun warms the Earth even if they have not even yet learned to read.

Yet now some Black Magicians have been casting a spell upon the world. These Black Magicians declare that the globe is getting hotter and hotter (even though it appears to be getting colder and colder) and that it is the fault of man for burning oil, coal and gas. And furthermore they allege that mankind is now causing the weather! They call it climatechange. So powerful are the spells of these Black Magicians that even highly intelligent men and women, and even some who opt to follow spiritual paths, have been taken in by this mass hysteria.

In the past, primitive man has worshipped the Sun or the spirits of Great Nature, with good cause, for these surely are the instruments of the Almighty. But now it appears and is even taught in schools that man is more powerful than Great Nature. So now man is blamed for extreme hurricanes, as happened in the Philippines recently. Of course the climate changers declare they are only talking about 'climate' not 'weather', but what is a hurricane other than a weather event? By extension of their logic, man must also be credited with the calm and sunlit weather. Will they next claim that man can manage the Jetstream? Or the orbit round the Sun?

The truth is – peace be to all the acolytes of Al Gore and all those Liberal members of Parliament who consider in their arrogance that it is their mandate to save The Planet – that Great Nature is infinitely more powerful than man. All talk of controlling climate and by extension weather is an affront to every man or woman who believes in the One Almighty God. And furthermore it is an affront to every scientist who adheres to the scientific method. These Black Magicians are propagandists not scientists in the true sense at all.

How many Wind Turbines are there already in Wales? At a rough count 146, with three times that number planned. Look it up on the Internet and discover just how big is each Wind Farm. Then multiply 146 x £ 100,000 = £ 14,600,000. That is 14 million pounds sterling, 600,000 pounds that is being poured out to rich landowners with no certainty that the wind will blow at the right force at the right time of demand.

By extension we can also look at the number of Turbines already in existence in England and Scotland to see the extent of the graft. Do you realise that there is even no certainty that the electricity that they might produce is even used? But it is paid for out of the public purse. That means out of the taxation that you and I pay. So we then pay twice, as our electricity bills explode. No wonder that David Cameron has at last woken up to the full extent of this scandal and has unwittingly admitted the need to get rid of this green crap.

Do we in England also need Wind Turbines that cost an arm and a leg, that are demonstrably intermittent, that are hugely inefficient, that are impoverishing the multitude, when in fact we already have the capacity, when we are shutting down the coal fired stations that we do have?

In the meantime, while our costs are rocketing, the price of power in the USA is falling rapidly through the extraction of shale gas. Germany has woken up to the fact that wind power is a non-starter and is building 25 new coal-fired Power Stations!

So let me ask you one final question, what is the driving force behind all this 'green crap'? Right in one – it is the material force.

Anthony Bright-Paul

Sunday, 01 December 2013

Wind Energy: How We Got Here

john droz, jr Physicist
  as at 12.06.2014

The first practical use of electricity (in the late 1800s), is generally attributed to Thomas Edison (a founder of GE). Of course there were actually dozens of other people who contributed to making commercial electricity a reality — and there were MANY formidable hurdles to overcome.

One of the initial primary issues was: *where was this electricity going to come from?* For the first **hundred± years**, there were six over-riding concerns about commercial electricity generators. Could they:

> **1** - provide *large amounts* of electricity?
> **2** - provide *reliable* and *predictable* electricity?
> **3** - provide *dispatchable*[1] electricity?
> **4** - service one or more *grid demand element*[2]? **5** - have a *compact*[3] facility?
> **6** - provide *economical* electricity?

[1] A *dispatchable* source generates power on a *human-defined* schedule.

[2] *Grid Demand Elements* = *Base Load* (minimum amount of electric power required 24/7) + *Load Following* (power output responds to moment-to-moment changes in system demand) + *Peak Load* (the maximum load during a specified time period).

[3] *Compact* is the ability to site an electrical facility on a relatively small and well-defined footprint, preferably near high demand, e.g. cities. This saves on expensive transmission lines, which can have significant power loss.

A primary goal of these efforts was to achieve **capacity**. To ensure reliability at the lowest cost, grid operators consider capacity in several ways as they evaluate electricity sources — but the most

important is *Capacity Value.* The layperson's definition of this is: "the percentage of a source's rated capacity that grid operators can be confidant will be available during future times of greatest demand."
Knowing this accurately is the key to reliable system grid performance!

Back to our history: several options were proposed to satisfy the above criteria. To maximize public benefit, each was individually and *scientifically* vetted to ascertain whether the suggested source would comply with ALL of the needed conditions.

Our careful implementation of these has resulted in the world's most successful grid system.

Over time, what resulted from these assessments was that we selected the following sources to provide commercial electricity: **hydroelectric**, **coal**, **nuclear**, **natural gas**, and **oil**. (Oil is the smallest source, supplying only 1±% of U.S. electricity.)

Note that each of these current sources meet **ALL** *of the above six essential criteria —* and if they don't **then they get replaced:** by conventional sources that **do** meet all criteria.

As a result, today**, and a hundred years from now**, these sources can provide ALL of the electrical needs of our society — *and continue to meet all six criteria.*

Note that **ALL** of the primary conventional sources use home-grown energy. Regarding our electrical energy sources, we have *always* been **energy independent!**

So what's the problem?

Ahhh, the problem is that a **new** element has been recently added to the list of requirements: *environmental impact.* The current number one environmental impact consideration is *greenhouse gas emissions* (e.g. CO_2).

Why has this joined the Big Six? It is a direct result of the current debate on Global Warming. Despite what the media conveys, this is not yet a scientifically resolved matter. In response to intense political pressure, our government has acquiesced to make emissions an *additional* criterion.

Having the government **mandate** that utility companies change the principles that have been the foundation of our electrical supply system for a hundred years — for reasons not yet scientifically resolved — is rather disconcerting.

And there's more. **Concern #3** is that this new standard for electrical supply sources now has taken priority *over ALL THE OTHER SIX!* In fact, this new-boy -on-the-block has in reality become the *ONLY* important benchmark — the other six are now given only lip service!

In this unraveling of sensibility there is one final incredible insult to Science: commercial electricity alternative sources that *claim* to make a consequential impact on CO_2, don't even have to prove that they actually do it!

Let's look at the environmental poster child: *wind energy*, and examine these criteria:

1 - Does industrial wind energy provide large amounts of electricity?

> **Yes, it could**. However, its effectiveness from most perspectives is inferior. For instance (because of the wide fluctuations of wind), on average, it produces only about 30% of it's nameplate power. Then compare these energy densities (MJ/kg): nuclear = 88,000,000, gas = 46, wind = .00006.

2 - Does industrial wind energy provide *reliable* and *predictable* electricity?

> **NO**. Despite the wind industry's best efforts it is not reliable or predictable *compared to the standards set by our conventional electrical sources*. What's worse is that when power is really needed (e.g. hot Summer afternoons) wind is usually on vacation.

3 - Does industrial wind energy provide *dispatchable* electricity?

> **NO**. Due to its unpredictability, wind can't be counted on to provide power *on-demand*, i.e. on a human-defined schedule.

4 - Does industrial wind energy provide one or more of the *grid demand elements*?

NO. It can not provide *Base Load* power, *Load Following,* or *Peak Load*. Essentially wind energy is just thrown into the mix and gets used who knows where.

5 - Is industrial wind energy *compact*?

NO. To even approximate the nameplate power of a conventional facility, like nuclear, takes over *a thousand times* the amount of area. Wind promoters try to convince non-technical politicians that it can have real capacity value. Their tinkertoy "solution" is to connect multiple wind farms spread over vast areas (often several states). In Australia it has been proven that this doesn't work. Even if it did, this would undermine the objective to be a *concentrated* power source.

Another "feature" of wind energy is that most of the windiest sites (and available land) are a LONG way from where the electricity is needed. This will result in *thousands* of miles of transmission towers and cables, at an *enormous* expense to ratepayers.

6 - Does industrial wind energy provide *economical* electricity?

NO. It is artificially subsidized WAY more than any conventional power source. A 2008 report by the US Energy Information Administration concluded that wind energy is subsidized to the tune of **$23** per megawatt-hour. By contrast, normal coal receives **44¢** per megawatt-hour, natural gas **25¢**, hydroelectric **67¢**, and nuclear power **$1.59**. [*Since these other sources meet ALL six criteria, there is some basis for subsidizing them!*]

And now the latest rule de jour:

7 - Does industrial wind energy make a *consequential reduction of CO_2*?

NO! No scientific study has ever proven that wind energy saves a meaningful amount of CO_2. In fact, the most scientific study done (by the *National Academies of Sciences*) says the opposite. Their 2007 report concludes that (assuming the *most optimistic conditions*) the U.S. CO2 savings by **2020** will amount to only **1.8%**. This is a trivial quantity, and amounts to about 1/80,000 of the world's CO2.

What about the critical factor of *Capacity Value*? The result of the above deficiencies is that wind energy has a Capacity Value of about **zero**. Compare this to the conventional sources, where essentially all of them have a Capacity Value near **100%**: *a stunning disparity.*

How can this *possibly* be? How could the U.S. be on the path to spend over a TRILLION dollars on an electrical source that **fails** five out of six of our historically important criteria, *AND has no scientific proof that it even meets this new emissions criterion?*

It's all about the money. Lobbyists for parties who want a piece of this TRILLION dollars, are leaving no stone unturned. Environmentalists who have taken their eye off the ball are promoting this palliative non-solution. Politicians eager to be seen as "green" are saying yes to everything the color of money.

Wind energy proponents typically try to rationalize away its serious shortcomings saying that things will "get worked out" *mañana*. What essentially is happening though, is that our politicians are trying to pound a square peg into a round hole. **Zero** wind energy is appropriate until **after** these significant problems are resolved — as some may *never* be (due to the laws of physicis).

After understanding wind energy's inherent electrical generation defects, it might put some other issues into perspective. For instance, it is entirely legitimate to be concerned about bird and bat mortality, noise intrusions, property devaluation, etc. But what if they were "fixed" (with a protective ordinance) — *would wind energy then be OK?*

NO: excellent regulations don't address the fundamental grid limitations of wind energy identified above. **Wind energy will not be acceptable until ALL seven criteria are met.**

Put another way: wind energy should not be allowed on the public grid until there is scientific proof that it is a net societal benefit.

Does wind energy's abysmal failure mean that all "renewables" are similarly deficient? **NO.** Each alternative power source should be scientifically evaluated. Industrial Geothermal holds significant promise. For scientifically based energy information, see **WiseEnergy.org.**

If we abandon our successful and time-tested criteria for selecting our sources of electrical power, and allow lobbyists to dictate our energy policies, there will be incalculable negative impacts on every person on the planet.

Can Air heat itself?

Anthony Bright-Paul

Monday, 17 March 2014

For that matter, can water heat itself? Can anything or can anybody heat themselves? It is a daft sort of question. Can an iron bar heat itself and form itself into a horseshoe? Not without a smithy and his forge. Can glass be made just from silica without the need for some 1700C of heat? No way – the raw materials of glass have to be heated to a high degree to make molten glass.

Fill a kettle with cold water from the tap. Can that water bring itself to the boil? Never in a million years. Will adding more water from the tap make any difference? Answer: None whatsoever.

Can the air heat itself? Can the atmosphere heat itself? Answer: There is no way that the air can heat itself. Yet the temperature of the air varies, as we are told with every Weather Forecast from the BBC.

In every case there has to be an agent. In the case of a kettle of water it must either be placed over a hot flame, of if electric, the current must be switched on. There is no other way.

If the waters of the oceans warm, there has to be an agent, and while hot springs rise up from the centre of the earth, without doubt the main agent is the Sun. As the Sun warms the oceans they in turn warm lower atmosphere by contact, that is, by conduction. This is the principal intermediary – radiation a minor player.

Let us look at the air temperature in a house. It is wintertime, so the living room is too cold for comfort. What is the answer? By no means will the room heat itself – the air cannot heat itself. So the central heating is switched on or a fire is lit. The air temperature in the room can only be modified by an agent – the air cannot modify itself.

Exactly the same principle applies to the air outside. The gases of which the air is composed cannot warm themselves. It is completely impossible for the Oxygen content to warm the Nitrogen, or the Water Vapour to warm the Methane. Even more it is completely impossible for a very minor constituent of the air, namely Carbon Dioxide, to warm the other constituents. Yet that is more or less what the Warmists propose.

I say 'more or less' since they do propose an agent in the first place, namely a low infrared that emanates from the Earth, which infrared has originally emanated from the Sun. But then they argue that this trace gas warms the rest of the atmosphere. But that is impossible, since CO_2 is just part of the atmosphere. The atmosphere cannot warm itself. There has to be an agent.

There has to be causation. There is a lot written about climate sensitivity. Yes, the gases of the atmosphere may be sensitive, but sensitivity must not be confused with causation. Carbon Dioxide is so sensitive that it can be frozen to make Dry Ice, or it can be liquefied, or it can be warmed. But it is not the agent. Carbon Dioxide cannot warm anything of itself, although it can be warmed.

Do the radiators in your living room warm your living room? Certainly not, if the central heating is not switched on. But if the central heating is switched on the agent is the boiler, or more exactly the combustion within the boiler. That heats the water that is pumped around the system. Actually the radiators do very little by way of radiating, as can be judged by holding your hand close to the radiator. However if you touch the radiator it maybe extremely hot to the touch. The Radiators should more correctly be called Conductors, since their surfaces warm the air by contact, that is, by conduction.

So much confusion is caused by the wrong or the sloppy use of language. It is true that every gas has its own heat capacity and it is true that everything above absolute zero will radiate its heat away, always from hot to cold, always from the higher energy to the lower. But what does the heating in the first place? What is it that generates heat? It is no good saying that climate sensitivity is the degree to which Carbon Dioxide causes a rise in atmospheric temperature, for the simple reason that Carbon Dioxide is not causal. This is not a question of science, but a question of the right use of language.

For a mere £15 it is possible to buy a tube of Carbon Dioxide over the Internet. I challenge any of the greatest Physicists to demonstrate its heat creating properties within the confined space of a room. Puff! Puff! Puff away! That Carbon Dioxide will never create, will never generate heat.

Sun heats the Earth and Earth warms the atmosphere – Yes, but only if the Sun warms the earth. The earth and the oceans are warmed by the Sun, so in reality it is the Sun that does all the warming.

So yes, there is Global Warming during the daytime – that is Solar Warming. It is impossible for anything other than the mighty Sun to warm the Planet Earth. The idea that Man could warm the Planet and cause a rise in temperature is a product not of science but of hallucination. Not science, No, but malfeasance or deliberate distortion.

But what about insulation? Sure insulation insulates. I have a conservatory facing south. On a clear frosty morning in winter the conservatory will warm up pleasantly and quite rapidly, as it is single glazed. But had it been double-glazed what would have been the result? It would have warmed up but much more slowly and likewise, once warmed it would have cooled down more slowly. This is very analogous to a dry or wet atmosphere. Since the incoming infrared from the Sun is far greater than the feebler infrared from the Earth, the atmosphere does more to keep us cool than to make us hot. This is the real and observable Greenhouse Effect.

Anthony Bright-Paul

Monday, 17 March 2014

Arse about Face

Anthony Bright-Paul

Saturday, 22 March 2014

In 2007, the APS improbably stepped out of the world of physics and into the world of policy and proclaimed:

> The evidence is incontrovertible: Global warming is occurring. If no mitigating actions are taken, significant disruptions in the Earth's physical and ecological systems, social systems, security and human health are likely to occur. We must reduce emissions of greenhouse gases beginning now.

The APS is the American Physical Society, which produced this statement in 2007, a statement that has been repeated by many other bodies since. Happily it is being challenged by three leading Skeptic scientists right now. You can read all about it in Climate Depot.

Some people questioned my article 'Can Air heat itself?' saying that nobody had ever made any such suggestion, that Air could heat itself. I beg to differ. Please read the above carefully. The clear implication is that emissions of Greenhouse Gases, a minor 1% of the air, were and are causing Global Warming. In fact this has been the battleground for the past 10 years and more.

We still have members of Parliament screaming that we must cut emissions of Carbon Dioxide. We still have people who blow themselves black and blue in their faces declaring, against all the evidence to the contrary, that the Globe is warming and that man is to blame for it.

Effectively the Warmists are saying that the Atmosphere warms itself. But if they are not saying that, if I have misunderstood them, then my friends, I must ask you, 'What causes a change in temperature?'

Last weekend the temperatures reached a balmy 19°C in the South East corner of England. This weekend coming the forecast maximum is 11°C. So I ask you again, 'What causes a change in temperature? Do the gases of the air change themselves? Or is there an outside agent or force that compels these changes?'

It is a very simple question and should not cause any anxiety or gnashing of teeth, as if I had let the whole side down, meaning the august body of Skeptic scientists. In fact I am persuaded that the apparent differences are largely semantic.

The difficulty is that we have been bombarded with lies for so long that it has become difficult to even question these lies, particularly if senior scientists and government advisers utter them. How often have I heard the words parroted 'Scientists say...' totally disregarding the fact that there are huge differences of opinion amongst scientists, totally disregarding the fact that there are good and bad scientists, totally disregarding the fact that many scientists have been corrupted by their remit.

So I want to ask the question in another form. Does the temperature drive the atmosphere or does the atmosphere drive the temperature? I am asking this question in a simple way that can be understood by any layman and even more easily by someone who has a very basic knowledge of Physics.

So what happens when a gas is warmed? If we were to imagine a cubic metre of air within a wire cage, what would happen to this air if it were warmed, so that its temperature rose? Would it still be contained within the confines of this wire cage, or virtual box as Dr Darko Butina avers?

Well even a non-scientist will know enough to realise that the gas will expand and rise up, some of the molecules will escape the confines of the cage, and the distance between the molecules will increase. The density will decrease, which is why the air is so thin on Mt Everest, and why the climbers need Oxygen.

> As the air rises up by convection and thins it also cools. That is curious, don't you think? The air rises because it is hot and yet at 1,000 feet up it has cooled by 2 degrees. Those of you who have

enquiring minds, that is, those of you who were born sceptics and can never stop asking questions, may ponder the assertion of the Astrophysicist James Peden – surely one of the most intelligent and gifted Physicists on this Planet – that there is no such thing as cold – there is only absence of heat.

Some of you may also ask yourselves, What is in the space between the molecules as they get farther and farther apart? This is a question that puzzled me and about which I wrote an earlier article, 'What's in the Space?'

I urge all of you who take the trouble to read my outpourings not to have knee-jerk reactions, but to adhere to observations and evidence – the basis of the scientific method – and even to question the Physics that some of you may have learned at school and college.

So I return to the question: Do the gas molecules drive the temperature, or is it the other way round – Arse about Face? In order that those of you who can read a simple paper on the question of "Gas Laws and Greenhouse Theory Or Back radiation? What Back Radiation?" Here are some quotes by Dr Darko Butina, who can explain his thesis far batter than can I..

> **The property of molecules that describes the relationship between the volume and number of molecules within that volume is called density. So, when temperature increases, the density of the starting virtual box decreases. In simplistic terms, when the air is warmed it expands and rises, while when it is cooled, the density increases, it becomes heavier and therefore it falls. The gas law applies to every known molecule in its gas phase without a single exemption!**

Conclusion – gas molecules of an open system are driven by temperature and it is physically impossible for gas molecules of the open system to control temperature in any shape or form.

Anthony Bright-Paul

Saturday, 22 March 2014

Why are they Lying?

All 'Solutions' Are Futile: US & UK Science Academies: Cutting all CO2 emissions would not stop global warming: 'Essentially irreversible on human timescales'

'The U.S. National Academy of Sciences and the UK's Royal Society: 'The current CO2-induced warming of Earth is therefore essentially irreversible on human timescales.' (Climate Depot)

Anthony Bright-Paul
08 March 2014

Why are they lying? Why are they lying in their back teeth?

While the Niagara Falls is frozen, while Chicago is suffering the coldest spell for 30 years, while the snow disrupts the whole of the United States with airlines cancelling flights and businesses disrupted everywhere, these two august bodies come out with a declaration that is so ridiculous that it defies common sense.

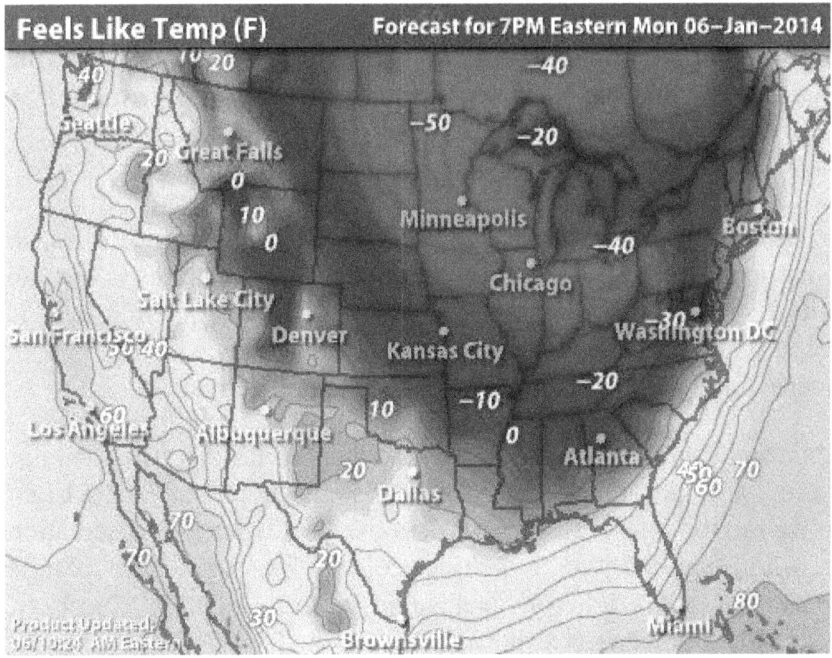

Firstly, is there any Global Warming at all? The facts are quite simple. There is nowhere, there is no one place, that one can take the temperature of the Earth. So they have to keep on repeating a lie, that Global Warming is continuing. Why do they repeat this absolutely unscientific assertion?

Secondly they assert that this non-existent and unproven warming is caused by CO2 emissions. Ask yourselves a simple question. What is supposed to be warming? Next ask your friends, particularly if they have bought this idea, ask them what is supposed to be warming?

It is a very simple question. Are the landmasses in the Northern Hemisphere getting hotter? Or the land masses in the Southern Hemisphere? Are the oceans getting hotter? And when you have gone through land and seas, then you can come to the atmosphere. **Is the Atmosphere getting hotter?**

Only a simpleton, only someone with no knowledge at all, could possibly declare that the atmosphere was getting hotter. Every weather bulletin in every country declares that the temperatures in their given locations are always changing, minute by minute. And that is only at ground level. Any Airline pilot can tell you the temperature at altitude is always less than at ground level. And that is to talk of just the Troposphere and to leave the Stratosphere, the Mesosphere and the Thermosphere as if they did not exist.

It is simply idiotic to talk of a Global temperature, since no such animal exists. And to take an average of air temperatures is also an exercise in fatuity. How can one take an average of something that is so vast and so diverse? These averages are subject to every sort of chicanery and manipulation and not to be relied on at all.

In hindsight only can we look back on major changes when ice has covered large portions of the Northern Hemisphere, and in hindsight can we see the intervening warm periods. To any normal observation the world appears to be getting colder. To argue that despite appearances the world on average is hotter is simply chicanery, duplicity and downright dishonesty.

It takes but a little thought to compare the hot with the cold. Can anyone seriously maintain that they would rather live in the frozen wastelands of the Arctic regions rather than the lush tropical paradises that abound, say in the Caribbean, say around the Mediterranean?

So why does this huge lie continue to exist? The answer is simple. Once there has been a political con of this magnitude there is no going back. Once people have been persuaded that Global Warming has been 'CO2 induced' with all the ramifications, with the huge cost and the false profits, those who have profited, those who have masterminded this huge confidence trick just cannot turn back. They would have to admit that they were wrong on a truly massive mind-boggling scale. That would mean all the scientists, all the politicians and all the ragtag and bobtail of the hysterical hangers-on would have to confess: We were all wrong.

Can you wonder that they hate the Skeptics who would expose them? The Warmists like to suggest that the Skeptics are in bed with Big Business, or with Big Oil or with Big Carbon, Big Coal. The truth is that the Warmists hate any form of Capitalism, yet love its benefits.

They hate any form of enterprise where individuals risk their capital on an enterprise for which they may make a profit – but also, let it not be forgotten, where they may also make a loss. In recent years many people invested in the Channel Tunnel. Was this a good investment? Only now after several years is this enterprise becoming profitable.

Consider the difference with Green Technologies like Wind Farms. These are intrinsically unreliable. Nobody with any sense would invest in a technology, which promised to produce intermittent electricity at 5 times the price of coal, oil or gas produced electricity. So these Technologies are inherently unprofitable and are subsidised by governments out of the Privy Purse. This means that a few rich landowners are paid by your and my money not to produce meaningful electricity. Do you wonder that some unscrupulous people are benefiting from what seems to be a freebie, a free handout from the State, when in reality the treasury is simply re-distributing our money to a few vassals in the form of monster bribes? Compared to that, Big Business is as white as the driven snow.

How can Carbon Dioxide, a gas, induce warming? Please note that weasel word 'induce'. The Warmists know very well that CO_2 cannot generate heat. Even by their own theory they only claim that:
-

> This light is mainly absorbed by land, sea etc and converted to heat (vibration of molecules) . When the energy is re-radiated it is emitted as much lower frequency infrared (below or lower frequency than red). It is this infrared radiation that is absorbed by the "greenhouse gases" (including water vapour), which warm the atmosphere. This is an oversimplification.

Here is an excellent simplification of the Warmist position, by my Warmist colleague Max Potter. Sunlight is converted to heat and the greenhouse gases absorb this lower frequency infrared. So what, even by their own theory, is doing the warming? It is the infrared, that is, the radiation coming from the earth that is absorbed by the molecules of Carbon Dioxide.

Sun heats Earth and Earth heats Atmosphere. What happens to warmed gases? They rise up by convection and cool. Apart from which the so-called Greenhouse gases comprise a mere 1% of the atmosphere, while 99% is Oxygen and Nitrogen, both of which gases have a higher heat capacity than Carbon Dioxide and furthermore are **transparent to radiation**. Not a lot of people seem to take that into account.

So what in reality warms the Lower Atmosphere? When the BBC gives a weather forecast and looks ahead at temperatures, do they say that the Carbon Dioxide will be 12 degrees today and the Oxygen and Nitrogen will be only 10 degrees! Don't be daft. The temperature that is given is the temperature of the air, of the atmosphere at a given moment, of all the gases together. But even the BBC will show a rise and a fall in temperatures from dawn till dusk and from dusk till dawn. Have you ever seen a temperature outside in the open remain absolutely constant?

The reality is that there is never-ending heat exchange between the surfaces of the land and oceans with the atmosphere **by conduction** all over the world. This heat exchange is immensely complicated, by winds, by clouds, by rain, and by the Jetstream coursing through the Stratosphere. Do you think these savants do not know this? Do you think that our Meteorologists are not aware of the reality?

The whole business of heat exchange by conduction is immensely complicated as different substances have different heat capacities.

But does CO2 do any warming at all? How can it? It is a gas. By the very words of my warmist colleague we can see that the miniscule amount of atmospheric CO2 (0.04% of the atmosphere) is warmed by this lower frequency infrared – or more exactly is absorbed and emitted. So what does the warming? The infrared. And where does the infrared come from? The infrared derives from the Sun's radiation. Carbon Dioxide does not and cannot generate heat and it can no more trap heat than can Nitrogen and Oxygen, which both have a greater heat capacity.

What we are concerned with here is the right use of language, with logic. Let us look at this warmist assertion again.

> It is this infrared radiation, which is absorbed by the "greenhouse gases" (including water vapour), which warm the atmosphere.

One needs to read this carefully. The Greenhouse Gases are part of the atmosphere. They are warmed by the infrared **and being gases they quickly lose their heat by convection. These gases do not do any warming – none whatsoever. Nor can they trap anything, least of all heat. It is completely impossible. And as to 'warming the atmosphere' this surely is the most unscientific statement ever. By what possible means could the molecules of a gas, which take their temperature from the surrounding molecules, by what possible means could they heat the other molecules of Nitrogen and Oxygen? By what possible means could they warm the atmosphere? By re-radiating downwards? If that were possible it would have been clearly demonstrated long before now. But there is no such demonstration.**

The Warmists love the phrase 'making the Lower Atmosphere warmer than it would otherwise be'. That's a scientific statement, if ever there was one.

What do we have from the witch doctors? Answer: An assertion that the Globe is warming. Where is the evidence? What precisely is warming? Nothing is clearly defined, is it? The very word 'Global' is suitably vague.

So now we have another assertion, namely, that this supposed warming is induced by Carbon Dioxide and, woe is me, cannot be stopped. Is this science? Or is this legerdemain?

There is only one source of heat for this Planet Earth and that is the Sun. Everybody in their hearts knows this for a fact. Everybody knows that there are seasons, which are governed by our journey round the Sun. Everybody knows than the weather changes from hour to hour and that the best meteorologists in the world cannot forecast more than a few hours ahead. The idea that mankind is somehow controlling the weather, and by

inference the climate; or the climate and by inference the weather, is not only ludicrous but is delusional.

So these Academies and academics keep repeating their mantras, keep repeating that the Globe is warming, keep repeating that that Carbon Dioxide is the culprit, keep repeating that mankind is changing the climate, all of which assertions have no scientific basis whatsoever. We are not dealing with scientists, but with magicians who have invoked a kind of mass suggestion, akin to hypnosis in order to lull the masses to sleep.

Millions of dollars and millions of pounds sterling have already been spent on an undefined and non-existent problem. Can you wonder that one lie has to be compounded with another? If they did not keep lying they would have to admit that they have been pouring away millions to no avail at all. Now they are trying to justify the waste they have created, a waste to no avail.

'Global warming' is rubbish says top professor

Emeritus Professor Les Woodcock
April 2014

'The term 'climate change' is meaningless. The Earth's climate has been changing since time immemorial that is since the Earth was formed 1,000 million years ago. The theory of 'man-made climate change' is an unsubstantiated hypothesis about our climate which says it has been adversely affected by the burning of fossil fuels in the last 100 years, causing the average temperature on the earth's surface to increase very slightly but with disastrous environmental consequences.

"The theory is that the CO2 emitted by burning fossil fuel is the 'greenhouse gas' that causes 'global warming' - in fact, water (vapour) is a much more powerful greenhouse gas and there is 20 time more of it in our atmosphere (around one per cent of the atmosphere) whereas CO2 is only 0.04 per cent.

"There is no reproducible scientific evidence CO2 has significantly increased in the last 100 years.

"Its absolutely stupid to blame floods on climate change, as I read the Prime Minister did recently. I don't blame the politicians in this case, however, I blame his so-called scientific advisors."

But surely most of the world's leaders, scientific community and people in general can't be wrong can they?

Prof Woodcock hits back: "This is not the way science works. If you tell me that you have a theory there is a teapot in orbit between the earth and the moon, its not up to me to prove it does not exist, its up to you to provide the reproducible scientific evidence for your theory.

"Such evidence for the man-made climate change theory has not been forthcoming."

He adds: "It's become almost an industry, as a consequence of this professional misconduct by Government advisors around the world, not just UK - you can't blame ordinary people with little or no science education for wanting to be seen to be good citizens who care about their grandchildren's future and the environment.

"In fact, the damage to our economy the climate change lobby is now costing us, is infinitely more destructive to the livelihoods of our grand-children. Indeed, we grand-parents are finding it increasingly expensive just to keep warm as a consequence of the idiotic decisions our politicians have taken in recent years about the green production of electricity.

"Carbon dioxide has been made out to be some kind of toxic gas but the truth is it's the gas of life. We breath it out, plants breath it in. The green lobby has created a do-good industry and it becomes a way of life, like a religion. I understand why people defend it when they have spent so long believing in it, people do not like to admit they have been wrong.

"In fact, the damage to our economy the climate change lobby is now costing us, is infinitely more destructive to the livelihoods of our grand-children. Indeed, we grand-parents are finding it increasingly expensive just to keep warm as a consequence of the idiotic decisions our politicians have taken in recent years about the green production of electricity.

"If you talk to real scientists who have no political interest, they will tell you there is nothing in global warming. It's an industry, which creates vast amounts of money for some people.

"Even the term 'global warming' does not mean anything unless you give it a time scale. The temperature of the earth has been going up and down for millions of years, if there are extremes, it's nothing to do with carbon dioxide in the atmosphere, it's not permanent and it's not caused by us. Global warming is nonsense."

(Excerpt published in Climate Depot of an article in the Yorkshire Evening Post, where Neil Hudson interviewed Professor Leslie Woodcock.)

THE TROUBLE WITH CLIMATE CHANGE

Nigel Lawson

27.05.2014

About the author: -
Lord Lawson of Blaby was Chancellor of the Exchequer from 1983–89 and served as Secretary of State for Energy from 1981–83. He is the author of several books, including Memoirs of a Tory Radical and An Appeal to Reason: A Cool Look at Global Warming. He is the chairman of the Global Warming Policy Foundation.

This essay has been slightly expanded from a speech given to the Institute for Sustainable Energy and the Environment at the University of Bath, the text of which was previously published in Standpoint magazine.

There is something odd about the global warming debate – or the climate change debate, as we are now expected to call it, since global warming has for the time being come to a halt.

I have never shied away from controversy, nor – for example, as Chancellor – worried about being unpopular if I believed that what I was saying and doing was in the public interest. But I have never in my life experienced the extremes of personal hostility, vituperation and vilification, which I – along with other dissenters, of course – have received for my views on global warming and global warming policies.

For example, according to Climate Change Secretary, Ed Davey, the global warming dissenters are, without exception, 'wilfully ignorant' and in the view of the Prince of Wales we are 'headless chickens'. Not that 'dissenter' is a term they use. We are regularly referred to as 'climate change deniers', a phrase deliberately designed to echo 'Holocaust denier' – as if questioning present policies and forecasts of the future is equivalent to casting malign doubt about a historical fact.

The heir to the throne and the minister are senior public figures, who watch their language. The abuse I received after appearing on the BBC's Today programme last February was far less restrained. Both the BBC and I received an orchestrated barrage of complaints to the effect that it was an outrage that I was allowed to discuss the issue on the programme at all. And even the Science and Technology Committee of the House of Commons shamefully joined the chorus of those who seek to suppress debate.

In fact, despite having written a thoroughly documented book about global warming more than five years ago, which happily became something of a bestseller, and having founded a think tank on the subject – the Global Warming Policy Foundation – the following year, and despite frequently being invited to appear on Today to discuss economic issues, this was the first time I had ever been asked to discuss climate change. I strongly suspect it will also be the last time.

The BBC received a well-organised deluge of complaints – some of them, inevitably, from those with a vested interest in renewable energy – accusing me, among other things, of being a geriatric retired politician and not a climate scientist, and so wholly unqualified to discuss the issue.

Perhaps, in passing, I should address the frequent accusation from those who violently object to any challenge to any aspect of the prevailing climate change doctrine, that the Global Warming Policy Foundation's non-disclosure of the names of our donors is proof that we are a thoroughly sinister organisation and a front for the fossil fuel industry.

As I have pointed out on a number of occasions, the Foundation's Board of Trustees decided, from the outset, that it would neither solicit nor accept any money from the energy industry or from anyone with a significant interest in the energy industry. And to those who are not – regrettably – prepared to accept my word, I would point out that among our trustees are a bishop of the Church of England, a former private secretary to the Queen, and a former head of the Civil Service. Anyone who imagines that we are all engaged in a conspiracy to lie is clearly in an advanced stage of paranoia.

The reason why we do not reveal the names of our donors, who are private citizens of a philanthropic disposition, is in fact pretty obvious. Were we to do so, they, too, would be likely to be subject to the vilification and abuse I mentioned earlier. And that is something which, understandably, they can do without.

That said, I must admit I am strongly tempted to agree that, since I am not a climate scientist, I should from now on remain silent on the subject – on the clear understanding, of course, that everyone else plays by the same rules. No more statements by Ed Davey, or indeed any other politician, including Ed Miliband, Lord Deben and Al Gore. Nothing more from the Prince of Wales, or from Lord Stern! What bliss!

Alarmism and its basis

But of course this is not going to happen. Nor should it; for at bottom this is not a scientific issue. That is to say, the issue is not climate change but climate change alarmism, and the hugely damaging policies that are advocated, and in some cases put in place, in its name. And alarmism is a feature not of the physical world, which is what climate scientists study, but of human behaviour; the province, in other words, of economists, historians, sociologists, psychologists and – dare I say it – politicians.

And en passant, the problem for dissenting politicians, and indeed for dissenting climate scientists, who certainly exist, is that dissent can be career-threatening. The advantage of being geriatric is that my career is behind me: there is nothing left to threaten.

But to return: the climate changes all the time, in different and unpredictable (certainly unpredicted) ways, and indeed often in different ways in different parts of the world. It always has done and no doubt it always will. The issue is whether that is a cause for alarm – and not just moderate alarm. According to the alarmists it is the greatest threat facing humankind today: far worse than any of the manifold evils we see around the globe which stem from what Burns called 'man's inhumanity to man'.

Climate change alarmism is a belief system, and needs to be evaluated as such. There is, indeed, an accepted scientific theory, which I do not dispute and which, the alarmists claim, justifies their belief and their alarm. This is the so-called greenhouse effect: the fact that the earth's atmosphere contains so-called greenhouse gases (of which water vapour is overwhelmingly the most important, but carbon dioxide is another) which, in effect, trap some of the heat we receive from the sun and prevent it from bouncing back into space.

Without the greenhouse effect, the planet would be so cold as to be uninhabitable. But, by burning fossil fuels – coal, oil and gas – we are increasing the amount of carbon dioxide in the atmosphere and thus, other things being equal, increasing the earth's temperature.

But four questions immediately arise, all of which need to be addressed, coolly and rationally. First, other things being equal, how much can increased atmospheric carbon dioxide be expected to warm the earth? (This is known to scientists as climate sensitivity, or sometimes the climate sensitivity of carbon.) This is highly uncertain, not least because clouds have an important role to play, and the science of clouds is little understood. Until recently, the majority opinion among climate scientists had been that clouds greatly amplify the basic greenhouse effect. But there is a significant minority, including some of the most eminent climate scientists, who strongly dispute this.

Second, are other things equal, anyway? We know that, over millennia, the temperature of the earth has varied a great deal, long before the arrival of fossil fuels. To take only the past thousand years, a thousand years ago we were benefiting from the so-called Medieval Warm Period, when temperatures are thought to have been at least as
warm, if not warmer, than they are today. And during the Baroque era we were grimly suffering the cold of the so-called Little Ice Age, when the Thames frequently froze in winter and substantial ice fairs were held on it, now immortalised in contemporary prints.

Third, even if the earth were to warm, so far from this necessarily being a cause for alarm, does it matter? It would, after all, be surprising if the planet were on a happy but precarious temperature knife-edge, from which any change in either direction would be a major disaster. In fact, we know that, if there were to be any future warming (and, for the reasons already given, 'if' is correct) there would be both benefits and what the economists call disbenefits. I shall discuss later where the balance might lie.

And fourth, to the extent that there is a problem, what should we, calmly and rationally, do about it?

Surface temperatures, past and projected
It is probably best to take the first two questions together. According to the temperature records kept by the UK Met Office (and other series are much the same), over the past 150 years (that is, from the very beginnings of the Industrial Revolution), mean global temperature has increased by a little under a degree centigrade – according to the Met Office, 0.8C. This has happened in fits and starts, which are not fully understood.

To begin with, to the extent that anyone noticed it, it was seen as a welcome and natural recovery from the rigours of the Little Ice Age. But the great bulk of it – 0.5C out of the 0.8C – occurred during the last quarter of the 20th century. It was then that global warming alarmism was born.

But since then, and wholly contrary to the expectations of the overwhelming majority of climate scientists, who confidently predicted that global warming would not merely continue but would accelerate, given the unprecedented growth of global carbon emissions as China's coal-based economy has grown by leaps and bounds, there has been no further warming at all. To be precise, the latest report of the Intergovernmental Panel on Climate Change (IPCC),1 a deeply flawed body whose non-scientist chairman is a committed climate alarmist, reckons that global warming has latterly been occurring at the rate of – wait for it – 0.05C per decade, plus or minus 0.1C.

Their figures, not mine. In other words, the observed rate of warming is less than the margin of error. And that margin of error, it must be said, is implausibly small. After all, calculating mean global temperature from the records of weather stations and maritime observations around the world, of varying quality, is a pretty heroic task in the first place. Not to mention the fact that there is a considerable difference between daytime and night-time temperatures. In any event, to produce a figure accurate to hundredths of a degree is palpably absurd.

The lessons of the unpredicted 15-year global temperature standstill (or hiatus as the IPCC calls it) are clear. In the first place, the so-called General Circulation Models which the climate science community uses to predict the global temperature increase which is likely to occur over the next 100 years are almost certainly mistaken, in that climate sensitivity is almost certainly significantly less than they once thought, and thus the models exaggerate the likely temperature rise over the next hundred years.

But the need for a rethink does not stop there. As the noted climate scientist Professor Judith Curry, chair of the School of Earth and Atmospheric Sciences at the Georgia Institute of Technology, recently observed in written testimony to the US Senate: -

> **Anthropogenic global warming is a proposed theory whose basic mechanism is well understood, but whose magnitude is highly uncertain. The growing evidence that climate models are too sensitive to CO2 has implications for the attribution of late-20th-century warming and projections of 21st-century climate. If the recent warming hiatus is caused by natural variability, then this raises the question as to what extent the warming between 1975 and 2000 can also be explained by natural climate variability.**

It is true that most members of the climate science establishment are reluctant to accept this, and argue that the missing heat has for the time being gone into the (very cold) ocean depths, only to be released later. This is, however, highly conjectural. Assessing the mean global temperature of the ocean depths is – unsurprisingly – even less reliable, by a long way, than the surface temperature record. And in any event most scientists reckon that it will take thousands of years for this 'missing heat' to be released to the surface.

In short, the effect of carbon dioxide on the earth's temperature is probably less than was previously thought, and other things – that is, natural variability and possibly solar influences – are relatively more significant than has hitherto been assumed. But let us assume that the global temperature hiatus does, at some point, come to an end, and a modest degree of global warming resumes. How much does this matter?

The question of impacts
The answer must be that it matters very little. There are plainly both advantages and disadvantages from a warmer temperature, and these will vary from region to region depending to some extent on the existing temperature in the region concerned. And it is helpful in this context that the climate scientists believe that the global warming they expect from increased atmospheric carbon dioxide will be greatest in the cold polar regions and least in the warm tropical regions, and will be greater at night than in the day, and greater in winter than in summer. Be that as it may, studies have clearly shown that, overall, the warming that the climate models are now predicting for most of this century is likely to do more good than harm.

This is particularly true in the case of human health, a rather important dimension of well-being. It is no accident that, if you look at migration for climate reasons in the world today, it is far easier to find those who choose to move to a warmer climate than those who choose to move to a colder climate. And it is well documented that excessive cold causes far more illnesses and deaths around the world than excessive warmth does.

The latest (2013–14) IPCC Assessment Report 3 does its best to ramp up the alarmism in a desperate, and almost certainly vain, attempt to scare the governments of the world into concluding a binding global decarbonisation agreement at the crunch UN climate conference due to be held in Paris next year. Yet a careful reading of the report shows that the evidence to justify the alarm simply isn't there.

On health, for example, it lamely concludes that 'the world-wide burden of human ill-health from climate change is relatively small compared with effects of other stressors and is not well quantified' – adding that so far as tropical diseases (which preoccupied earlier IPCC reports) are concerned, 'Concerns over large increases in vector-borne diseases such as dengue as a result of rising temperatures are unfounded and unsupported by the scientific literature'.

Moreover, the IPCC conspicuously fails to take proper account of what is almost certainly far and away the most important dimension of the health issue. And that is, quite simply, that the biggest health risk in the world today, particularly of course in the developing world, is poverty.

We use fossil fuels not because we love them, or because we are in thrall to the multinational oil companies, but simply because they provide far and away the cheapest source of large-scale energy, and will continue to do so, no doubt not forever, but for the foreseeable future. And using the cheapest source of energy means achieving the fastest practicable rate of economic development, and thus the fastest elimination of poverty in the developing world. In a nutshell, and on balance, global warming is good for you.

The IPCC does its best to contest this by claiming that warming is bad for food production: in its own words, 'negative impacts of climate change on crop yields have been more common than positive impacts'. But not only does it fail to acknowledge that the main negative impact on crop yields has been not climate change but climate change policy, as farmland has been turned over to the production of biofuels rather than food crops. It also understates the net benefit for food production from the warming it expects to occur, in two distinct ways.

In the first place, it explicitly takes no account of any future developments in bioengineering and genetic modification, which are likely to enable farmers to plant crops that are drought-resistant and which thrive at warmer temperatures, should these occur.

Second, and equally important, it takes no account whatever of another effect of increased atmospheric carbon dioxide, and one which is more certain and better documented than the warming effect, namely, the stimulus to plant growth: what the scientists call the 'fertilisation effect'. Over the past 30 years or so, the earth has become observably greener, and this has even affected most parts of the Sahel. It is generally agreed that a major contributor to this change has been the growth in atmospheric carbon dioxide from the burning of fossil fuels.

This should not come as a surprise. Biologists have always known that carbon dioxide is essential for plant growth, and of course without plants there would be very little animal life, and no human life, on the planet. The climate alarmists have done their best to obscure this basic scientific truth by insisting on describing carbon emissions as 'pollution' – which, whether or not they warm the planet, they most certainly are not – and deliberately mislabelling forms of energy which produce these emissions as 'dirty'.

In the same way, they like to label renewable energy as 'clean', seemingly oblivious to the fact that by far the largest source of renewable energy in the world today is biomass, and in particular the burning of dung, which is the major source of indoor pollution in the developing world and is reckoned to cause at least a million deaths a year.

Compared with the likely benefits to both human health and food production from CO_2- induced global warming, the possible disadvantages from, say, a slight increase in either the frequency or the intensity of extreme weather events is very small beer. It is, in fact, still uncertain whether there is any impact on extreme weather events as a result of warming (increased carbon emissions, which have certainly occurred, cannot on their own affect the weather: it is only warming which might). The unusual persistence of heavy rainfall over the UK during February, which led to considerable flooding, is believed by scientists to have been caused by the wayward behaviour of the Jetstream; and there is no credible scientific theory that links this behaviour to the fact that the earth's surface is some 0.8C warmer than it was 150 years ago.

That has not stopped some climate scientists, such as the publicity-hungry chief scientist at the UK Met Office, Dame Julia Slingo, from telling the media that it is likely that 'climate change' (by which they mean warming) is partly to blame. Usually, however, the climate scientists take refuge in the weasel words that any topical extreme weather event – whatever the extreme weather may be, whether the recent UK rainfall or last year's typhoon in the Philippines – 'is consistent with what we would expect from climate change'.

So what? It is also consistent with the theory that it is a punishment from the Almighty for our sins (the prevailing explanation of extreme weather events throughout most of human history). But that does not mean that there is the slightest truth in it. Indeed, it would be helpful if the climate scientists would tell us what weather pattern would not be consistent with the current climate orthodoxy. If they cannot do so, then we would do well to recall the important insight of Karl Popper – that any theory that is incapable of falsification cannot be considered scientific.

Moreover, as the latest IPCC report makes clear, careful studies have shown that, while extreme weather events such as floods, droughts and tropical storms have always occurred, overall there has been no increase in either their frequency or their severity.

That may, of course, be because there has so far been very little global warming indeed: the fear is the possible consequences of what is projected to lie ahead of us. And even in climate science, cause has to precede effect: it is impossible for future warming to affect events in the present.

Of course, it doesn't seem like that. Partly because of sensitivity to the climate change doctrine, and partly simply as a result of the explosion of global communications, we are far more aware of extreme weather events around the world than we used to be. And it is perfectly true that many more people are affected by extreme weather events than ever before. But that is simply because of the great growth in world population: there are many more people around. It is also true, as the insurance companies like to point out, that there has been a great increase in the damage caused by extreme weather events. But that is simply because, just as there are more people around, so there is more property around to be damaged.

The fact remains that the most careful empirical studies show that, so far at least, there has been no perceptible increase, globally, in either the number or the severity of extreme weather events. And, as a happy coda, these studies also show that, thanks to scientific and material progress, there has been a massive reduction, worldwide, in deaths from extreme weather events.

Scientific standards
It is relevant to note at this point that there is an important distinction between science and scientists. I have the greatest respect for science, whose development has transformed the world for the better. But scientists are no better and no worse than anyone else. There are good scientists and there are bad scientists. Many scientists are outstanding people, working long hours to produce important results. They must be frustrated that political activists then turn those results into propaganda. Yet they dare not speak out for fear of losing their funding.

Indeed, a case can be made for the proposition that today's climate science establishment is betraying science itself. During the period justly known as the Enlightenment, science achieved the breakthroughs which have so benefited us all by rejecting the claims of authority – which at that time largely meant the authority of the church – and adopting an overarching scepticism, insisting that our understanding of the external world must be based exclusively on observation and empirical investigation. Yet today all too many climate scientists, in particular in the UK, come close to claiming that they need to be respected as the voice of authority on the subject
– the very claim that was once the province of the church.

If I have been critical of the latest IPCC report, let me add that it is in many respects a significant improvement on its predecessors. It explicitly concedes, for example, that 'climate change may be beneficial for moderate climate change' – and moderate climate change is all that it expects to see for the rest of this century – and that 'Estimates for the aggregate economic impact of climate change are relatively small...For most economic sectors, the impact of climate change will be small relative to the impacts of other drivers.' So much for the unique existential planetary threat.

What it conspicuously fails to do, however, is to make any assessment of the unequivocally adverse economic impact of the decarbonisation policy it continues to advocate, which (if implemented) would be far worse than any adverse impact from global warming.

Even here, however, the new report concedes for the first time that the most important response to the threat of climate change must be how mankind has responded throughout the ages, namely intelligent adaptation. Indeed, the 'impacts' section of the latest report is explicitly entitled 'Impacts, Adaptation and Vulnerability'. In previous IPCC reports adaptation was scarcely referred to at all, and then only dismissively.

The importance of adaptation
This leads directly to the last of my four questions. To the extent that there is a problem, what should we, calmly and rationally, do about it?

The answer is – or should be – a no-brainer: adapt. I mentioned earlier that a resumption of global warming, should it occur (and of course it might) would bring both benefits and costs. The sensible course is clearly to pocket the benefits while seeking to minimise the costs. And that is all the more so since the costs, should they arise, will not be anything new: they will merely be the slight exacerbation of problems that have always afflicted mankind.

Like the weather, for example – whether we are talking about rainfall and flooding (or droughts for that matter) in the UK, or hurricanes and typhoons in the tropics. The weather has always varied, and it always will. There have always been extremes, and there always will be. That being so, it clearly makes sense to make ourselves more resilient and robust in the face of extreme weather events, whether or not there is a slight increase in the frequency or severity of such events.

This means, in the UK, measures such as flood defences and sea defences, together with water storage to minimise the adverse effects of drought; and in the tropics better storm warnings, the building of levees, and more robust construction.

The same is equally true in the field of health. Tropical diseases – and malaria is frequently (if inaccurately) mentioned in this context – are a mortal menace in much of the developing world. It clearly makes sense to seek to eradicate these diseases – and in the case of malaria (which used to be endemic in Europe) we know perfectly well how to do it – whether or not warming might lead to an increase in the incidence of such diseases.

And the same applies to all the other possible adverse consequences of global warming. Moreover, this makes sense whatever the cause of any future warming – whether it is man-made or natural. Happily too, as economies grow and technology develops, our ability to adapt successfully to any problems which warming may bring steadily increases.

Yet, astonishingly, this is not the course on which our leaders in the Western world generally, and the UK in particular, have embarked. They have decided that what we must do, at inordinate cost, is prevent the possibility (as they see it) of any further warming by abandoning the use of fossil fuels.

Even if this were attainable – a big 'if', which I will discuss later – there is no way in which this could be remotely cost-effective. The cost to the world economy of moving from relatively cheap and reliable energy to much more expensive and much less reliable forms of energy – so-called renewables, on which we had to rely before we were liberated by the fossil-fuel-driven Industrial Revolution – far exceeds any conceivable benefit.

It is true that the notorious Stern Review, widely promoted by a British prime minister with something of a messiah complex and an undoubted talent for PR, sought to demonstrate the reverse, and has become a bible for the economically illiterate. But Stern's dodgy economics have been comprehensively demolished by the most distinguished economists on both sides of the Atlantic. So much so, in fact, that Lord Stern himself has been driven to complain that it is all the fault of the computer models used, which – and I quote him – 'come close to assuming directly that the impacts and costs will be modest, and close to excluding the possibility of catastrophic outcomes'.

It may well be the case that these elaborate models are scarcely worth the computer code they are written in, and certainly the divergence between model predictions and empirical observations has become ever wider. Nevertheless, it is a bit rich for Stern now to complain about them, when they remain the gospel of the climate science establishment in general and of the IPCC in particular.

But Stern is right in this sense: unless you assume that we may be heading for a CO_2 - induced planetary catastrophe, a view for which there is no scientific basis, a policy of decarbonisation cannot possibly make sense.

A similar, if slightly more sophisticated, case for current policies has been put forward by a distinctly better economist than Stern, Harvard's Professor Martin Weitzman, in what he likes to call his 'dismal theorem'. After demolishing Stern's cost–benefit analysis, he concludes that Stern is in fact right but for the wrong reasons.

According to Weitzman, this is an area where cost–benefit analysis does not apply. Climate science is highly uncertain, and a catastrophic outcome, which might even threaten the continuation of human life on this planet, cannot be entirely ruled out, however unlikely it may be. It is therefore incumbent on us to do whatever we can, regardless of cost, to prevent this.

This is an extreme case of what is usually termed 'the precautionary principle'. I have often thought that the most important use of the precautionary principle is against the precautionary principle itself, since it can all too readily lead to absurd policy prescriptions.

In this case, a moment's reflection would remind us that there are a number of possible catastrophes, many of them less unlikely than that caused by runaway warming, and all of them capable of occurring considerably sooner than the catastrophe feared by

Weitzman; and there is no way we can afford the cost of unlimited spending to reduce the likelihood of all of them.

In particular, there is the risk that the earth may enter a new ice age. This was the fear expressed by the well-known astronomer Sir Fred Hoyle in his book Ice: The Ultimate Human Catastrophe, and there are several climate scientists today, particularly in Russia, concerned about this. It would be difficult, to say the least, to devote unlimited sums to both cooling and warming the planet at the same time.

At the end of the day, this comes down to judgment. Weitzman is clearly entitled to his, but I doubt if it is widely shared; and if the public were aware that it was on this slender basis that the entire case for current policies rested I would be surprised if they would have much support. Rightly so.

The global dimension
But there is another problem. Unlike intelligent adaptation to any warming that might occur – which in any case will mean different things in different regions of the world, and which requires no global agreement – decarbonisation can make no sense whatever in the absence of a global agreement. And there is no chance of any meaningful agreement being concluded. The very limited Kyoto accord of 1997 has come to an end; and although there is the declared intention of concluding a much more ambitious successor, with a UN-sponsored conference in Paris next year at which it is planned that this should happen, nothing of any significance is remotely likely.

And the reason is clear. For the developing world, the overriding priority is economic growth: improving the living standards of the people, which means among other things making full use of the cheapest available source of energy: fossil fuels.

The position of China, the largest of all the developing countries and the world's biggest (and fastest growing) emitter of carbon dioxide, is crucial. For very good reasons, there is no way that China is going to accept a binding limitation on its emissions.

China has an overwhelmingly coal-based energy sector – indeed it has been building new coal-fired power stations at the rate of one a week – and although it is now rapidly developing its substantial indigenous shale gas resources (another fossil fuel), its renewable energy industry, both wind and solar, is essentially for export to the developed world.

It is true that China is planning to reduce its so-called 'carbon intensity' quite substantially by 2020. But there is a world of difference between the sensible objective of using fossil fuels more efficiently, which is what this means, and the foolish policy of abandoning fossil fuels, which it has no intention of doing. China's total carbon emissions are projected to carry on rising – and rising substantially – as its economy grows.

This puts into perspective the UK's commitment, under the Climate Change Act, to near-total decarbonisation. The UK accounts for less than 2% of global emissions; indeed, its total emissions are less than the annual increase in China's. Never mind, says Lord Deben, chairman of the government-appointed Climate Change Committee, we are in the business of setting an example to the world.

No doubt this sort of thing goes down well at meetings of the faithful, and enables him and them to feel good. But there is little point in setting an example, at great cost, if no one is going to follow it; and around the world governments are now gradually watering down or even abandoning their decarbonisation ambitions. Indeed, it is even worse than that. Since the UK has abandoned the idea of having an energy policy in favour of having a decarbonisation policy, there is a growing risk that, before very long, our generating capacity will be inadequate to meet our energy needs. If so, we shall be setting an example all right: an example of what not to do.

Unreason and morality
So how is it that much of the Western world, and this country in particular, has succumbed to the self-harming collective madness that is climate change orthodoxy? It is difficult to escape the conclusion that climate change orthodoxy has in effect become a substitute religion, attended by all the intolerant zealotry that has so often marred religion in the past, and in some places still does so today.

Throughout the Western world, the two creeds that used to vie for popular support – Christianity and the atheistic belief system of Communism – are each clearly in decline. Yet people still feel the need both for the comfort and for the transcendent values that religion can provide. It is the quasi-religion of green alarmism and global salvationism, of which the climate change dogma is the prime example that has filled the vacuum, with reasoned questioning of its mantras regarded as little short of sacrilege.

The parallel goes deeper. As I mentioned earlier, throughout the ages the weather has been an important part of the religious narrative. In primitive societies it was customary for extreme weather events to be explained as punishment from the gods for the sins of the people; and there is no shortage of this theme in the Bible, either – particularly, but not exclusively, in the Old Testament. The contemporary version is that, as a result of heedless industrialisation within a framework of materialistic capitalism, we have directly (albeit not deliberately) perverted the weather, and will duly receive our comeuppance.

There is another aspect, too, which may account for the appeal of this so-called explanation. Throughout the ages, something deep in man's psyche has made him receptive to apocalyptic warnings that the end of the world is nigh. And almost all of us, whether we like it or not, are imbued with feelings of guilt and a sense of sin. How much less uncomfortable it is, how much more convenient, to divert attention away from our individual sins and reasons to feel guilty, and to sublimate them in collective guilt and collective sin.

Why does this matter? It matters, and matters a great deal, on two quite separate grounds. The first is that it has gone a long way towards ushering in a new age of unreason. It is a cruel irony that, while it was science which, more than anything else, was able by its great achievements to establish the age of reason, it is all too many climate scientists and their hangers-on who have become the high priests of a new age of unreason.

But what moves me most is that the policies invoked in its name are grossly immoral.

We have, in the UK, devised the most blatant transfer of wealth from the poor to the rich – and I am slightly surprised that it is so strongly supported by those who consider themselves to be the tribunes of the people and politically on the Left. I refer to our system of heavily subsidising wealthy landlords to have wind farms on their land, so that the poor can be supplied with one of the most expensive forms of electricity known to man.

This is also, of course, inflicting increasing damage on the British economy, to no useful purpose whatever. More serious morally, because it is on a much larger scale, is the perverse intergenerational transfer of wealth implied by orthodox climate change policies. It is not much in dispute that future generations – those yet unborn – will be far wealthier than those – ourselves, our children, and for many of us our grandchildren – alive today. This is the inevitable consequence of the projected economic growth which, on a 'business as usual' basis, drives the increased carbon emissions that in turn determine the projected future warming. It is surely perverse to abandon what is far and away the cheapest source of energy in order that future generations avoid any disadvantages that any warming might bring: this simply impoverishes those alive today in order to ensure that future generations, who will be signally better off regardless of

what happens today, are better off still.

However, the greatest immorality of all concerns those in the developing world. It is excellent that, in so many parts of the developing world – the so-called emerging economies – economic growth is now firmly on the march, as they belatedly put in place the sort of economic policy framework that brought prosperity to the Western world. Inevitably, they already account for, and will increasingly account for, the lion's share of global carbon emissions.

But, despite their success, there are still hundreds of millions of people in these countries in dire poverty, suffering all the ills that this brings, in terms of malnutrition, preventable disease, and premature death. Asking these countries to abandon the cheapest available sources of energy is, at the very least, asking them to delay the conquest of malnutrition, to perpetuate the incidence of preventable disease, and to increase the numbers of premature deaths.

Global warming orthodoxy is not merely irrational. It is wicked.

(Copyright 2014 The Global Warming Policy Foundation.)

Everywhere and All-at-Once

Anthony Bright-Paul
Friday, 25 October 2013

Once I had begun to take in the full import of the fact that the earth and the oceans meet the atmosphere just everywhere and all-at-once I began to search for a word to describe this phenomenon that everybody observes yet which virtually nobody acknowledges. Could one say simultaneous-ness? Ugly. Then I thought of 'synchronicity' – but it is not enough. Synchronous applies only to time. I needed a word for time and place.

The great heat exchange that is taking place over the entire globe is taking place everywhere and all at once by contact, by conduction. There is not a word in the English language that I am aware of that can describe the fact that is easily observable - that the action and reaction that is taking place between every surface everywhere on this planet is instantaneous, contiguous and continuous.

Far from it being the Sceptics who do not believe in Global Warming and Climate Change, precisely the opposite is the case. It is the Warmists who do not acknowledge, who are utterly blind to the incredible GlobalWarmingCooling that is taking place all together and all at once – in a sort of majestic symphony. Every instrument in this celestial orchestra is playing its part harmoniously, everything is on cue, and the supreme conductor, Great Nature, is conducting everything.

As I rose yesterday morning to clear the misted windows of my car, the temperature was a mere 7C. I had to drive with my wife down the M3 past Southampton, along the M27 and then the A31 across the New Forest. As I passed the sign for the Rufus Stone and came to the great open spaces of heath and gorse, I was reminded that only a few minutes ago in geological time William the Conqueror was hunting here. And being full of the wonder of the contact between the heavens and the earth just everywhere and all at once, I sensed the atmosphere of this wild place. Then, later, as we passed along the A31 towards Wimborne, high pines with their own secret atmospheres bordered the road. Everywhere ahead of me the ribbon of road unwound, now warmer at some 17C, unusually mild for late October. The road itself was an immense heat exchange mechanism. Elsewhere in the world the sun is so hot in some places that the road tar melts and yet elsewhere again the ice-truckers drive their perilous way in the frozen wastes.

A good friend wrote to me that he was concerned about the incremental increase in world temperatures – only a little increase will make a huge difference, he averred. (Incremental? Where did he get that word from?) Tell me good friend, where have you observed this incremental increase? And if it does pertain, have you taken into account the incremental increase in Antarctic ice? Even the Arctic ice, which ebbs and flows, is presently at an all time high. Al Gore suggested it might all be gone by Christmas – he forgot that during the Holocene Maximum there was almost certainly no ice in the Arctic at all! For three thousand years!

Oh no! Good friend, you have been listening to the querulous whingeing sounds of the warmists' tin whistle. You have forgotten the winds, the woodwind section, the imperious sound of the horns, the clarion call of the brass, the thunder of the drums, the lightning clash of the symbols and the sunny serenity of the strings.

It is no good to isolate one tiny element like Carbon Dioxide and then imagine that you can pontificate about the great body of the Biosphere. We need a holistic approach. It is incumbent on those who seek the truth to listen to the entire orchestra that is playing everywhere and all at once.

My friend the truly honourable Emmanuel Elliott emailed me, insisting that I listen to some Feedback programme on the BBC, - Should Climate Sceptics be given Airtime? This included some short comments from Professor Bob Carter, who happened to have been my second mentor, and who went out of his way to visit me in my home, when over from Australia. After a minute and a half at most of Bob Carter the Feedback droned on for all of 11 minutes, justifying the fact that the BBC are shit-scared to debate the Climate issues.

Why? They argued that climate scientists were all agreed on the facts, so actually there was nothing to discuss – the consensus had it.

But of course this was a barefaced lie, delivered with cool aplomb. What is true and what is accepted by both sides of the argument is that Carbon Dioxide absorbs and emits infrared radiation. The whole of the warmist position is based on this fact, established in experiments many moons ago. But the Warmists proceed to a *non sequitur*, a classical *non sequitur*.

They argue that through the warming of the Carbon Dioxide in the air, the whole atmosphere is warmed; and the more Carbon Dioxide the more warming will result. But the empirical facts deny this, even the so-called facts from the Climatic Research Unit. But to argue that the warming and cooling of CO_2, which is just 0.04% of the atmosphere, is going to warm the whole of the rest of the atmosphere, calls for an abandonment of all logic and the curt dismissal of the scientific method.

Once we establish that the principal heat exchange mechanism is through contact, through 'conduction' the arguments of the Warmists melt into nothing. Radiation may be absorbed and emitted by a few pesky molecules of Carbon Dioxide, but how does that compare with the all embracing contact of Earth and Atmosphere Everywhere and All-at-Once?

If you want crap science

Anthony Bright-Paul
26/02/2014

If you want crap science then look no further than Google and look up the causes of Global Warming, or man-made, or Anthropogenic Warming. Website after website is designed for dummies, that is for those who are gullible, suggestible and even more so for those who are hysterical.

Let us think for one moment about what we can all certainly agree on. Is there anybody alive who doubts or who can doubt the Sun warms the Earth? It is so self- evident that it is often overlooked. It is evident that the heat from the Sun is greatest at the Equator and the tropics and is least at the Poles. This is so elementary that I feel diffident even to put it into writing. This is simply caused by the angle of the Sun upon our spinning globe.

Secondly everyone is agreed that seasons are a fact of experience, that while it is winter in the Northern Hemisphere it is summer in the Southern. Neither does anyone dispute that we orbit the Sun in an ellipse, which means that there are times when we are nearer the Sun and times when we are farther, by as much as 3, 4, or 5 million miles. We are swinging round the Sun at a rate of 67,062 miles per hour. Everyone also agrees that the Earth is spinning on its own axis, which accounts for daytime and nighttime. And everyone has agreed to this since the days of Copernicus, Galileo and Giordano Bruno, though these great gentlemen had to suffer every sort of indignity and humiliation before their ideas were accepted. Copernicus conveniently died before his great work was published; Galileo was forced to make a confession of guilt under threat of torture on the rack; and the great monk Bruno was held in prison for seven years before being burnt alive at the stake.

So we owe these great scientists a huge debt. And it is not meet, it is not right, that we should allow the scientific truths for which they suffered at the hands of ignoramuses to be disbanded and cast aside, in favour of a scientific mumbo-jumbo pursued by a small cabal of so-called climate scientists, who could more readily be called quacks.

Have I any justification for making this assertion? Let us therefore examine what the Warmists, those who insist that man is warming the Globe; let us examine what they have to say, whether it is true, whether it is false, or whether it is full of half-truths.

So let us start with 'What is Global Warming?'
http://usliberals.about.com/od/environmentalconcerns/a/GlobalWarm1.htm
Which was the first site that appeared when I Googled the above question.

> "The global surface temperature is an estimate of the global mean surface air temperature. However, for changes over time, only anomalies... are used, most commonly based on the area-weighted global average of the sea surface temperature anomaly and land surface air temperature anomaly."

You have to read that first sentence carefully to understand what it means. It is an **estimate** of the global mean surface air temperature. So it is not the temperature of the air surface for the very good reason that the surface is changing second by second everywhere upon this Planet. Not only that, these supposed temperatures are based on what is called 'anomalies'. Let me put this in layman language. They are based on averages against former pre-supposed averages and they can be fiddled just any way the Warmists want, if they are so disposed. If you want anomalies this is where you get them.

> "The Intergovernmental Panel on Climate Change (IPCC) concludes that... **greenhouse gases** are responsible for most of the observed temperature increase since the middle of the twentieth century, and that natural phenomena such as solar variation and volcanoes probably had a small warming effect from pre-industrial times to 1950 and a small cooling effect afterward.

Hallo! Just read this second paragraph carefully. Let me ask you, dear reader, have you observed personally, have you observed a temperature increase where you live? I mean was the temperature around your home static, and has that that was static increased? Codswallop. The temperatures that we all enjoy are going up and down as the sun rises and the sun sets and according to the seasons. A gas cannot increase its own temperature! The atmosphere cannot heat itself.

Don't be taken in by scientific gobbledegook. **greenhouse gases are responsible for most of the observed temperature increase** I am afraid that that is utter rubbish. It is the sort of rubbish that is designed to fool the common man. In the first place there is not any observed increase, and neither can a Greenhouse Gas be responsible for any supposed increase. Why? I will quote from a correspondence I had with Dr Helen Czerski, who far from being a Sceptic is a Warmist, despite her excellent Television series on "Orbit: Earth's Journey Round the Sun."

> Of course the temperature of carbon dioxide varies - it is the same temperature as the other gases around it. And the atmosphere has different temperatures at different heights, largely dictated by the adiabatic lapse rate.

So how can a gas whose temperature varies warm anything? The simple answer is that it cannot. A gas by law cannot generate heat. Helen Czerski went on: -

> The point is that carbon dioxide absorbs and *re-emits* energy. So it doesn't heat up, but it gives the energy away almost immediately, and because it re-emits the energy in all directions, some of this re-emitted energy goes downwards. This energy is re-absorbed by what's below, making the lower part of the atmosphere hotter than it would otherwise be.

Here she enunciates clearly the classical warmist position, and it is exceedingly interesting the way she puts it. According to this renowned Physicist the Carbon Dioxide molecules in the atmosphere do not warm up, but give their energy away almost immediately, making the lower part of the atmosphere hotter than it would otherwise be.

But that final assertion is impossible. The source cannot re-heat itself. Since the lower part of the atmosphere is the warmest, as per the aforesaid Adiabatic Lapse Rate, a higher and colder part cannot possibly heat what is warmer below. If that were possible then there would be empirical evidence, but there is none and a little thought shows just why that cannot be.

Dr Phillip Bratby, who is a PhD physicist and has worked all his life in industry, doing all sorts of heat transfer calculations, (being in industry, his calculations had to be correct and accurate) put it most succinctly in direct answer to Dr Helen Czerzki: -

> Certainly, radiation is emitted by all molecules at all temperatures and in all directions, but there is only net radiation (hence heat flow) from hot molecules to cold molecules. As long as there is a temperature gradient, heat is transferred by radiation upwards through the atmosphere, none is transferred downwards by radiation." In none of my calculations did I need to consider 'back-radiation' as it is always cancelled out in the Stefan Boltzman T^4 difference.

So Dr Bratby agrees that radiation is emitted by all molecules every which way including earthwards, but **there is only net radiation from hot to cold molecules.**

Let us return now to the Global Warming website.

> Atmospheric radiation is emitted to all sides, including downward to the Earth's surface. Thus, greenhouse gases trap heat within the surface-troposphere system. This is called the greenhouse effect."

I am afraid this sort of pap is taught to schoolchildren. Back Radiation is a defiance of the 2nd Law of Thermodynamics and as for the Greenhouse Effect it has long been blown apart.

To quote now Dr Darko Butina in his paper 'Gas Laws and Greenhouse Theory' "Every molecule has the ability to absorb heat, a property quantified by its heat capacity." He goes on: - "Please note that a statement like "CO2 has the capacity to trap heat is totally meaningless since every molecule has the ability to absorb or trap heat."

> Conclusion – *gas molecules of an open system are driven by temperatures and it is physically impossible for gas molecules of the open system to control temperature in any shape or form.*

The atmosphere is an open system, not a closed box. There is no igloo in the sky!

The truth of the matter is that the Warmists with their obsession with Carbon Dioxide have got everything topsy-turvy. Far from the gases of the atmosphere being responsible for warming the atmosphere, precisely the opposite is the case. Without the atmospheres the Earth would be as hot as Hades by day and as cold as the Moon is by night. Our atmosphere mitigates the heat of the sun by day and releases the heat by night.

> When sunlight hits the moon's surface, the temperature can reach 253 degrees F . F (123 C). The "dark side of the moon" can have temperatures dipping to minus 243 F (minus 153 C).

http://www.space.com/18175-moon-temperature.html

Certainly a humid atmosphere can slow the exit of heat more than a dry one, but that is all.

The obsession with 'man-made' warming is a sinister political shibboleth having little to do with real science. It is maintained by corrupted rather than corrupt scientists, since they are corrupted by their very remit – to find a human fingerprint. That is, they begin with a conclusion before assembling the facts or evidence.

Surely only if the Earth were knocked off its Orbit and flew nearer the Sun, only in that way could the Earth be made substantially warmer than it is.

The end of phoney science is near. When Michael Mann the creator of the famed Hockey Stick graph that Al Gore used in his film, when Michael Mann was unable to produce his metadata in a court case against Professor Tim Ball, this was surely the beginning of the end of the greatest scam since the South Sea Bubble.

The collapse of the Warmists is nigh, though they will put up a desperate rear guard fight as is happening right now.

Anthony Bright-Paul

Tuesday, 25 February 2014

Night and Day

Anthony Bright-Paul
Saturday, 10 May 2014

It is a curious fact that amongst our most learned scientists, who can use symbols and do all sorts of arcane mathematics that most of us unlearned cannot remotely understand, there are many who take no account of one most obvious fact that is in the experience of every man, woman and every child upon this planet. And that is that, as the sun sets, night begins and as the sun rises in the East day begins.

Even that is also curious, for no matter where you live, whether in New Zealand or Alaska, whether on the Russian steppes or in Bechuanaland, the sun always rises in the East.

One can hardly blame the ancients who only a few minutes ago in geological time sincerely believed that the Earth was the centre of the universe. Truly speaking, as a child, did you believe anything else? The sun rose and the sun set – surely we must be in the middle of the universe. But we know now that Copernicus was lucky to die before his great work was published, that Galileo only just escaped being put on the rack until his arms were pulled out of their sockets, and Giordano Bruno was not so lucky, as that brave monk endured seven years in jail before he was burnt at the stake. Yet nowadays Dr Helen Czerski was able to make a documentary to great acclaim, showing the Earth's Orbit round the sun.

Even so many "scientists" still behave or theorise as if there were no night and no day. Now my Warmist friend actually makes a very good explanation as far as it goes.

> *Most of the energy radiated by the sun is in the band of the electromagnetic spectrum we see as light, because the Sun is very hot ("white hot"). This is not absorbed by any of the gases in the air (it is absorbed by soot or dust particles but that is a different matter). This light is mainly absorbed by land, sea etc and converted to heat (vibration of molecules). When the energy is reradiated it is emitted as much lower*

> *frequency infrared (below or lower frequency than red). It is this infrared radiation that is absorbed by the "greenhouse gases" (including water vapour), which warm the atmosphere*
> Max Potter email 4/11/2012

Not the atmosphere, Max! The infrared radiation is absorbed by the Greenhouse gases at ground level, but not by the atmosphere, for he has just proven, as he clearly says 'This is not absorbed by any gases in the air'. Of course, what he means is that the incoming radiation from the sun is not absorbed by Oxygen and Nitrogen, which are transparent to radiation, thank the Good Lord. But by the same token the energy that is re-radiated is not and cannot be absorbed by the air either, only by the Greenhouse Gases. If Oxygen and Nitrogen are transparent to near-IR then they must also be transparent to far-IR. The Greenhouse gases at ground level may absorb the lower frequency radiation, but this will leave the 99% of the air untouched. And since the Greenhouse gases are also subject to convection they cool rapidly as they rise. Any back or downward radiation cannot possibly warm the lower atmosphere, which is by his own account **transparent to radiation**.

If we look at the diagram of the incoming light from the sun we see clearly that it has the whole spectrum from ultraviolet to infrared. Yes, he is correct to say that N2 and O2 do not absorb the incoming radiation, but what about water? We can see water up in the atmosphere, because water has many forms. The most obvious forms of water and water vapour are clouds, steam, fog, mist and so on. Water in the atmosphere clearly does absorb incoming radiation. It just so happens that today is a typical Spring day, with a strong west wind and a mixture of sunshine and showers. Can there be any doubt about it, since I and my wife have sat outside one moment with our resident vixen, who has become amazingly tame, and two magpies trotting about on our lawn, while the next moment as the clouds obscure the sun the temperature drops abruptly and we are forced indoors as droplets of water rain down.

It is interesting that water in the atmosphere is both visually opaque and opaque to radiation, while those great scientists, John Tyndall and Svante Arrhenius have demonstrated that Carbon Dioxide, although transparent, is also opaque to radiation.

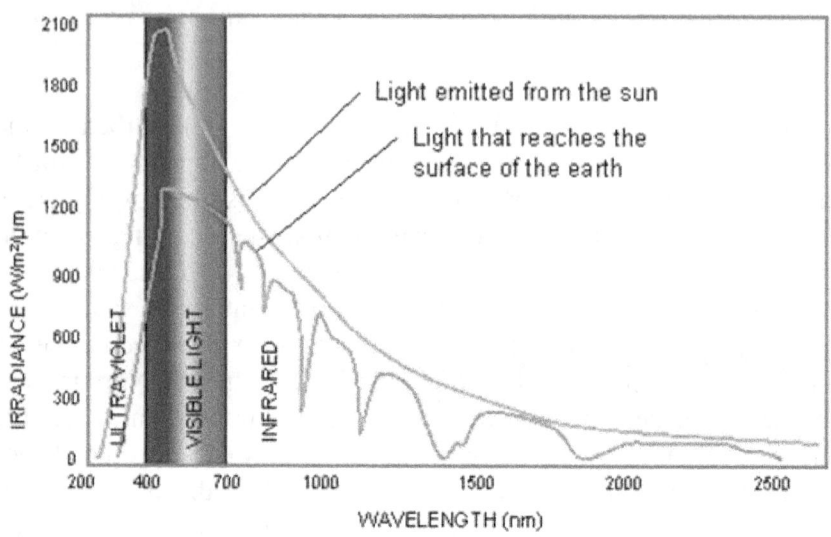

So it is clear that far from the Greenhouse Gases causing any warming whatsoever, precisely the opposite is the case. Nobody has to be a scientist to observe the truth of this matter, but only to be a sentient being, someone who is awake and not befuddled by drink, drugs or any other incapacities, such as preconceived ideas. And this is absolutely true and can hardly be challenged during the daytime. This is best explained in Hans Schreuder's paper 'Greenhouse Gases in the atmosphere cool the Earth', which paper is No 34 in this book

Why, during the day? The answer to that must be obvious. During the day the earth and we who are upon it are receiving the sun's insolation. As the Sun warms the Earth and seas, so they in turn warm the atmosphere. Any sentient being can also experience a warm balmy breeze, and intense Saharan heat and or a cold wind. How can that be? As Dr Darko Butina makes clear in his paper on Gas laws, O_2 and N_2 have a higher heat capacity than CO_2.

Max, with his excitation of molecules, actually gives the answer. The atmosphere is warmed by conduction from the earth, which includes the waters and oceans. Hence we have what is called the Adiabatic Lapse rate, for, as earth warms the gases, they rise and cool by convection.

We must remember that the idea that emissions of Carbon Dioxide are causing Global Warming is just a theory. A lot of people have taken it as an established scientific fact. But this is not only far from the case, it is just an unsupported theory, that has become a sort of religious belief for some people.

Just as the Greenhouse Gases scatter the incoming infrared, so also do these same gases absorb, emit and delay the heat that is leaving the earth by convection and by this lower frequency infrared. We all know that as night falls the warmth of the day will decline slowly in a humid atmosphere, as in Jakarta, but will decline rapidly in the dry Saharan atmosphere. The Greenhouse gases do not and cannot possibly cause any Global Warming whatsoever. The one and only causative factor is the sun. Not one of the gases can generate heat, but they can all be warmed and they can all be cooled. Happily Max Potter has also agreed that the gases of the atmosphere do not generate heat - of course, since the gases can be warmed or cooled or frozen, or even liquefied.

As he makes clear, radiation has to encounter mass to generate heat. That is one way to generate heat – then there is combustion, friction, fusion, fission and so on. So the gases themselves are passive.

The Sun is some 93 million miles away – Max will know the exact figure. As our orbit round the Sun is elliptical it is difficult to be exact. Between our planet and the Sun is outer space. Since our great scientists have calculated that the corona of the Sun to be some 6,500°C how hot then is Outer Space? Is it unimaginably hot or is it unimaginably cold? Well, Max will tell you I am sure, that it is neither. It is neither hot nor cold – it has no temperature. Why? Because outer space is a vacuum. It is empty. As the saying goes 'you can't heat nothing.'

How does one measure Global Warming? Since any column of air will decline by 2°C for every 1,000 feet of altitude, at what height above sea level do you take the temperature?

Sunday, 11 May 2014

So let us try to sum up in simple language comprehensible to any layman what our friend Max Potter has to say.

> The gases of the air are transparent to radiation. I will qualify that and say that 99% of the air that is Nitrogen and Oxygen is transparent to radiation.
> Dust and other particles do absorb incoming radiation. May I add water vapour, clouds and also Carbon Dioxide absorb and scatter this insolation.
> The incoming radiation is absorbed by land and sea and converted to heat (vibration of molecules). As radiation encounters mass, agreed.
> When the energy is re-radiated it is emitted as a low frequency infrared. Not quite. The vast amount of the heat is directly transmitted by conduction to the air, which is why and how the Oxygen and Nitrogen warm up.
> It is this infrared radiation that is absorbed by the "greenhouse gases" (including water vapour), which warm the atmosphere. Half correct. The infrared radiation is absorbed by the greenhouse gases at ground level, which then rise by convection and cool. However since Max has already argued that the atmosphere is transparent to radiation, the atmosphere cannot possibly be warmed by a lower frequency radiation if it was transparent to incoming radiation. I am sure that Max on reflection will agree.

Monday, 12 May 2014

As the sun rises in the morning we are witnesses to the great heat exchange. All through the night even as we sleep we have been breathing in oxygen and exhaling carbon dioxide. Not only has every living man and woman been exhaling this carbon dioxide, but also so has every beetle, every spider, every squirrel, every fox and all the bacteria and every single animal in the animal kingdom – they have all been inhaling oxygen and exhaling carbon dioxide.

Everywhere the vegetable kingdom are rejoicing as this wonderful transparent gas Carbon Dioxide is for them breakfast, lunch and dinner. By a wonderful conjuring trick that is called photosynthesis green plants absorb this carbon dioxide hungrily and convert part of it back into oxygen. There is a very simple experiment that one can do with a couple of glass jars, one on the kitchen window and one below the sink, which any man can do in order to see how green leaves produce oxygen for us as a by-product. Nowadays it is possible to buy tomatoes all the year round, as massive greenhouses are built with climate control, where carbon dioxide is pumped in to feed these hungry plants.

As the sun rises and lights up approximately half the planet, the great heat exchange takes places. As the radiation from the sun travels through the transparent gases of the atmosphere (as our friend Max Potter has pointed out) a massive excitation of molecules takes place on the dry and liquid surfaces, over an enormous area of land and sea, producing heat. The thrashing seas warm up, releasing carbon dioxide into the air like fizzy champagne. Where the sun is directly overhead, as over the Sahara, the sand that before dawn was as cold as stone, warms up and as it warms these hot sands warm the atmosphere by contact, by touching, by conduction.

The atmosphere is like a huge mantle touching every surface, of sea, of lakes, of rivers, of rock, of sand, of tarmac and every blade of grass and every single leaf on every tree. The atmosphere, which is 99% nitrogen and oxygen, which is transparent to radiation, now warms up by conduction. The heated air then rises and cools as it rises. Great masses of hot air are blown by the winds, so that we in the south of England often experience hot winds, sometimes laden with sand, coming off the Sahara.

Every substance has its own heat capacity and this can easily be tested by touch. As the sunlight reaches the roof of your car the cold metal warms, the interior warms and the dashboard. Put your hands on them each and severally to experience the different heat capacities. In the garden the wooden seat has one heat capacity and the concrete paving slabs another. The grass warms up, and the dew evaporates, that is it turns into water vapour, with silvery wraiths that are visible to the eye. Water vapour is opaque.

So we see here the distinctive features of the Greenhouse Gases, which are opaque to radiation, while Nitrogen and Oxygen are transparent to this same radiation. The great heat exchange that takes places everywhere and all at once is primarily a matter of touch, of conduction. The low-grade radiation from the earth during the night, which the Philistine scientists make much of, is a minor player in this great heat exchange. So our friend Max Potter is right in much that he says, but has he drawn the right conclusions from his own data?

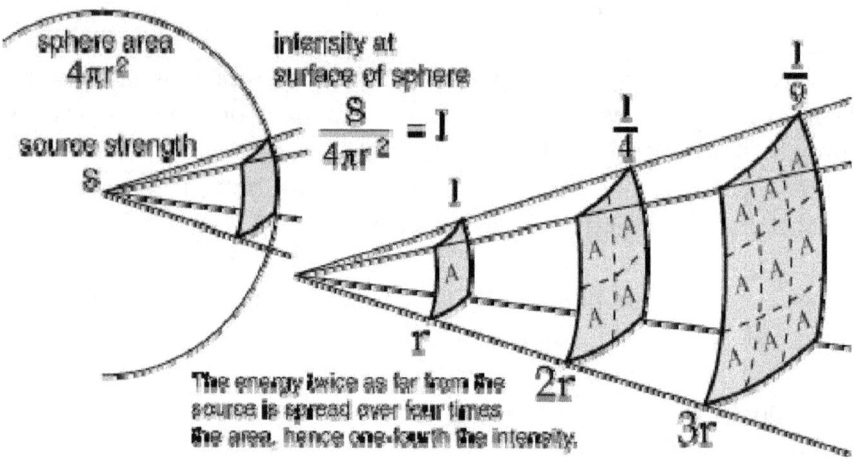

Radiation is subject to a law that is called the Inverse Square law. As the radiation spreads out from its source it decreases in intensity. This can easily be tested in what one of my friends calls my kitchen physics. Bring a kettle to the boil. Hover your hands close to the kettle to feel the heat radiation. Now clasp the kettle with both hands but release quickly. In this simple experiment you can experience at first hands (excuse the pun) the difference between radiation and conduction. The transfer of heat by conduction is immediate, whereas the transfer by radiation is relatively slow and diminishes rapidly with distance. The same experiment can be experienced with a bonfire, where the radiant heat may be extreme five feet way, but is negligible at fifteen feet.

The Philistine scientists, who are bloodsuckers and parasites upon the public purse, make much of so-called man-made Global Warming, as if Carbon Dioxide causes or even could cause warming. The air does not warm us; it is we and the surfaces that warm the air. It is easy to be confused because it is pleasant to live in a warm atmosphere. It is not always easy to see that the atmosphere warms from the earth or seas upwards. However it is easy to see the great white cap on Mt Blanc. It is easy to experience the difference in heat even at the relatively low altitudes of the Lake or Peak districts. Anyone who flies a plane whether in England or say California will immediately experience the fact that heat declines by 2°C for every 1,000 feet of altitude.

While smoke from fossil fuels will contain Carbon Dioxide, these Philistine and lying scientists seek deliberately to confuse the average layman, by suggesting by inference that smoke itself is Carbon Dioxide, and that Carbon Dioxide is a pollution and must be controlled or captured and put underground, in this way selling their souls for a mess of pottage, which is the Biblical way of saying a government grant from the public purse. We still get these Philistine scientists, who should and do know better (Al Gore and even our own Met Office) talking about Carbon pollution and advising such innocents as President Obama and in the UK David Cameron, who listen only to those whose advice fits in with their political perspectives. Every one of these so-called scientists seek to confuse the public and above all to deny to enquiring and sceptical scientists a platform. This is a heinous sin wherever it occurs.

As Lord Lawson has pointed out in his recent Bath lecture (which will also be included in my book in its entirety) the consequent cost to the population of Britain, the United States, Australian and Canada is enormous. To quote his final paragraphs:
.
> However, the greatest immorality of all concerns the masses in the developing world. It is excellent that, in so many parts of the developing world "the so-called emerging economies" economic growth is now firmly on the march, as they belatedly put in place the sort of economic policy framework that brought prosperity to the Western world. Inevitably, they already account for, and

the incidence of preventable disease, and to increase the numbers will increasingly account for, the lion's share of global carbon emissions.

But, despite their success, there are still hundreds of millions of people in these countries in dire poverty, suffering all the ills that this brings, in terms of malnutrition, preventable disease, and premature death. Asking these countries to abandon the cheapest available sources of energy is, at the very least, asking them to delay the conquest of malnutrition, to perpetuate of premature deaths.

Global warming orthodoxy is not merely irrational. It is wicked.

Issued through the Global Warming Policy Foundation
10 Upper Bank Street, London, London E14 5NP
www.thegwpf.org

I really liked the ending of this piece, as it is the first time I have ever seen anyone state that the Warmist position is wicked. Of course when Lord Lawson says '...the lion's share of global carbon emissions' he means carbon dioxide emissions. But as per the argument in the foregoing essay the fear that is spread that carbon dioxide will or can produce global warming is in fact groundless.

Once my friend, Max Potter, takes in the true import of his own piece quoted, he will realise he has destroyed his own warmist arguments. Similarly those Sceptics who refer to Climate sensitivity are, I believe, only conceding that Carbon Dioxide does absorb and is opaque to infrared radiation. But once the full import of the transparency of Nitrogen and Oxygen to either near–IR or far-IR is fully apprehended, then at one blow both the GHT, Greenhouse Theory, and Climate sensitivity are put to bed. This is the death knell of Anthropogenic Global Warming.

Solar Power Distribution

Anthony Bright-Paul
17.05.2014

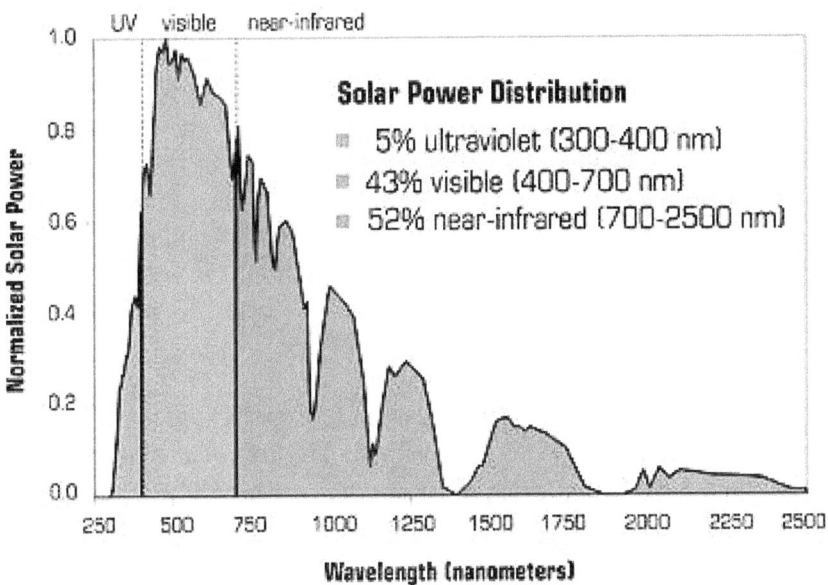

Our friend, Max Potter, has expressed his theory of how the atmosphere works and I hope to share his view with you that he wrote to me in an email many moons ago.
So here is Max's explanation:

> *Most of the energy radiated by the sun is in the band of the electromagnetic spectrum we see as light, because the Sun is very hot ("white hot"). This is not absorbed by any of the gases in the air (it is absorbed by soot or dust particles but that is a different matter). This light is mainly absorbed by land, sea etc and converted to heat (vibration of molecules). When the energy is reradiated it is emitted as much lower frequency infrared (below or lower frequency than red). It is this infrared radiation that is absorbed by the "greenhouse gases" (including water vapour), which warm the atmosphere*

Max Potter email 4/11/2012

Let us take his first sentence, which is very good as far as it goes. As we can see by the diagram above, 5% is ultraviolet, 43% is visible light and 52% is near infrared. (Near-IR is the scientific way of expressing incoming infrared.). So that is the Solar Power Distribution, which Max has this very day agreed.

Max goes on: This is not absorbed by any of the gases in the air (it is absorbed by soot or dust particles but that is a different matter). Here Max makes a very important point. As we all know the atmosphere is composed of 99% Nitrogen and Oxygen and 1% of the Greenhouse Gases, Water Vapour, Carbon Dioxide and Methane etc. So it is indeed remarkable that that 99% is transparent to radiation, as Max makes clear. However it is not correct to say that it is not absorbed by the Greenhouse Gases and scattered, as it is certainly scattered and reflected by clouds, which are Water Vapour.

Just imagine if Oxygen and Nitrogen were not transparent, the whole temperature gradient would be reversed. The top of the Troposphere would be the hottest, as the atmosphere would be hot from the top down. In fact if Oxygen and Nitrogen were not transparent to radiation life on Earth would be impossible.

On the other hand infrared radiation is absorbed by Water Vapour and Carbon Dioxide as Max points out later. Of course, it is absorbed and emitted by both near-IR and far- IR, scattering and dispersing the near-IR. Otherwise we would be as hot as the Moon by day and as cold as the Moon by night. (see Article 42 Living on the Moon.)

This light is mainly absorbed by land, sea etc and converted to heat (vibration of molecules). Max clearly refers to the whole incoming solar power, but he is right – it is absorbed by the land and the sea and is converted to heat. This is why the atmosphere is warmed from the bottom upwards. This is a concept that is not easy to grasp first of all. It can be expressed as, radiation has to encounter mass to produce heat, so it is very lucky for us that Nitrogen and Oxygen are transparent to radiation, otherwise we would be living in a fiery furnace, that is, if life on Earth could exist at all.

When we sit out in the sun it is not always easy to realise that as we are 'mass' it is the Sun's radiation that makes us hot and we are heating the atmosphere not the other way round. Of course the air gets heated by conduction from the earth and from us. If the atmosphere gets hot it becomes difficult for us to keep cool, as it inhibits heat loss. So we perspire, as evaporation helps to keep us cool.

Now what happens to hot air? Imagine that you are in the middle of the Sahara with David Attenborough. The sand has reached 70°C and the air temperature close to the ground is 41°C. As the air rises it cools by 2°C for every 1,000 feet of altitude. Comes sunset and the Sun goes down. What happens to all that hot air? As it rises up and cooler air comes in it gets cold enough that the Bedouin have to light fires.

Similarly we can see hot air balloons. Filled with hot air they rise up into the atmosphere. Gliders roam about seeking thermals, hot air rising from the ground. So we do not have to be a scientist to know that hot air rises – we have only to watch a bonfire on November 5th.

When the Sun goes down the warmed earth and seas release their heat by convection and hot air rises. Now Max adds an important point. The Earth also radiates its heat away at night.

> *It is this infrared radiation that is absorbed by the "greenhouse gases" (including water vapour), which warm the atmosphere.*

That is a non-sequitur. No! The Greenhouse Gases absorb and scatter both Near-IR and Far-IR and far from 'warming' have a net cooling effect. (Go back to article 34. Greenhouse Gases in the Atmosphere cool the Earth).

The Earth and the Oceans warm the gases of the atmosphere primarily by contact, by conduction, which is immediate. The infrared may be absorbed and emitted by the 1% of the atmosphere that is Greenhouse Gas, but it cannot warm the atmosphere, since 99% is transparent to infrared, and is being warmed by conduction in any case. As gases are warmed they rise up and cool by convection.

This supposed warming of the atmosphere is a classic misconception of the Warmists and cannot be demonstrated or proven in any way, for the simple reason that it is false.

Today has been gloriously hot, 23°C at 1PM, but it was muggy and close, lots of big clouds and water vapour in the air. Not till after 5PM could I make myself mow the lawn. As the hot air was rising the temperature dropped and a cool breeze arrived.

However with all this humidity the night is likely to be warm, as the temperature will subside slowly, thanks to the Greenhouse Gases. They do such a wonderful job of keeping us cool in the day and delaying the exit of hot air in the night.

So what Max makes clear is that the Earth, that is, the oceans, the rivers, the lakes, the deserts, the forests, the buildings, the roads are all heated by Solar Power – his 'excitation of molecules'. The atmosphere – that is N2 and O2 - is transparent to radiation, but is subject to conduction and convection. Hot air rises!

And that is why the balloons go up.

Heat and a Thermos Flask

Anthony Bright-Paul
Saturday, 12 July 2014

Our mutual friend Dave Haskell wrote to me once again concerning 'trapping heat' and unfortunately being away from home a couple of days I read this email on my Galaxy Ace and inadvertently erased it. However I remember the contents well enough to declare that on matters of fact I believe that the two of us are in complete agreement – the only difference being our understanding of the word 'trapped'.

For Dave if the sun heats the interior of a car and the temperature is raised, the heat is trapped. Similarly with an oven or a greenhouse. Even more so with a thermos flask, since a thermos can keep coffee, say, hot for a long time.

Dave also agrees when the sun goes down the interior of a car will cool all by itself, so what may be very hot on a sunny afternoon will have cooled off by the next morning. So we are agreed on all of this. We are also agreed that a thermos flask is the most efficient way of preserving heat, but we also agree that it is not 100% efficient.

So our only difference is in semantics – that is in the way we use the word 'trapped'. It is his prerogative to use the word 'trapped' in the way that he does.

Say we fill a thermos flask before a journey at 8AM with boiling water and coffee grains at 100°C and stop for lunch at 1 o'clock; the coffee will still be so hot that we need to add milk. However, imagine that the day is so hot that we buy some cold drinks to eat with our sandwiches and the thermos flask is unopened until the following morning for breakfast. Ugh! By this time the coffee is tepid and undrinkable, and fit only to be poured down the drain.

When I use the word 'trapped' I refer to a completely steady state. Heat is trapped only if I put in 100 degrees and get out 100 degrees.

However I am sure that Dave will agree that such a state of efficiency does not exist.

Heat is either being generated or being dissipated.

Heat can be generated by friction, by fusion, by fission, by compression (as Dave points out) and by combustion. On the other hand all heat <u>by itself</u> always flows from hot to cold and never vice-versa.

Everyone agrees that in a confined space as in an oven if heat is generated then temperatures will rise. That fact is not in dispute. However our atmosphere is not a confined space. As the radiation from the Sun encounters the mass of our earth and oceans, heat is generated. At nighttime when heat is no longer being generated, cooling takes place as warmed airs rise and are displaced by cooler airs.

Greenhouse gases can only delay the entrance and the exit of heat, they cannot generate heat and therefore cannot make anything hotter. Certainly Carbon Dioxide cannot generate heat, as even my most formidable Warmist opponents have agreed. They imagine that heat is trapped, but that of course is impossible – heat is either being generated or being dissipated. There is no steady state.

In my book then, a substance, say milk, can be trapped in a bottle. Hot water may be trapped in a hot water bottle. Hot coffee may be trapped in a thermos flask. They are all substances. But 'heat' is not a substance – Heat is a genie! Heat is a Houdini! Heat is never trapped, imprisoned, or incarcerated. Heat will always escape!

Anthony Bright-Paul
Saturday, 12 July 2014

The Sand Lizard

Anthony Bright-Paul
Friday, 09 May 2014

I was so intrigued by one small moment in David Attenborough's documentary 'Sahara' that I had to find it and play it again on my computer. Just before the amusing incident with the dung beetle, there is a picture of this sand lizard on a pile of twigs with his head as far up as he could get it, in the air, and Attenborough remarks that
in this midday heat of 41°C the lizard was getting some relief from the heat, even though he was only centimetres from the ground. And to be reminded that the sands, were some 71° C

And he was absolutely correct. A few centimetres up he would be marginally cooler than the temperature immediately above the ground. Of course, had the sand lizard been able to fly 1,000 feet up the temperature would have dropped from 41° C to 39°C. Had he been able to fly 5,000 feet up the temperature would have been a mere 31°C. And the temperature would have gone on decreasing by 2°C for every 1,000 feet of altitude.

But in the middle of the Sahara, with the midday sun beating down remorselessly, it was amazing to be reminded that the temperature of the atmosphere declines with altitude. In those stark and dry conditions it seemed scarcely feasible. Why was it not hotter centimetres up? Indeed why was it not hotter still nearer that blazing sun? In fact why was it not like a fiery furnace the closer you got to the sun?

Here we have a wonderful illustration and proof of the fact that the sun heats the earth and the earth heats the atmosphere. But although my mentor, my guru in matters scientific, had insisted that this was so, I still had my lingering doubts. Was he absolutely correct? Had not Galileo said that the atmosphere has mass? Does it not have more mass where the molecules are most closely packed? OK, if radiation has to encounter mass to produce heat why does the air not get hot from above? Why does it get hot from the bottom up?

While the Warmists concentrate their attention on the Greenhouse Gases they overlook the truly remarkable attributes and heat capacity of Nitrogen and Oxygen. While the Greenhouse Gases respond to the various forms of heat exchange, namely conduction, convection and radiation, Nitrogen and Oxygen, which make up 99% of the atmosphere, are **transparent to radiation**.

Just imagine if this were not so. At the present moment we all know, or should know, about Standard Atmosphere, which is used by airline pilots all over the world. We know that wherever you are on the surface of this planet, the temperature of the atmosphere, at least the Troposphere, declines by 2°Celsius for very 1,000 feet of altitude. However if N2 and O2 were not transparent to radiation, then the whole temperature gradient would be reversed. At the Karman line on the edge of space some 62 miles up, I have been assured by the astrophysicist James Peden, that the very few molecules that are up there, as it is still near vacuum, can be as hot as 1,500°C. Imagine that, percolating all the way down to the surface. We would be living, if life could exist at all, within a fiery furnace.

What David Attenborough has shown, by an almost throwaway line in the middle of a documentary, is something very intriguing. The atmosphere in the midday sun in the middle of the Sahara is undoubtedly hot. I say undoubtedly, because the heat can be measured by a thermometer. And since this is an exceptionally dry and arid area of the planet we have to ask ourselves, what warms the atmosphere?

If the midday air in the Sahara reaches 50°C it can only be through conduction. Attenborough has, albeit inadvertently, illustrated precisely how the atmosphere is warmed. And he has also illustrated precisely how it cannot be warmed. So it may well be true that Carbon Dioxide and Water Vapour molecules absorb and emit heat, and that Carbon Dioxide radiates the heat that it has absorbed every which way including downwards, but it makes no difference. Why? Because 99% of the atmosphere, which is Nitrogen and Oxygen, is transparent to radiation.

Could Carbon Dioxide then warm the other molecules of the atmosphere by conduction? Well, a single hot molecule might just warm a colder molecule as per the 1st and 2nd laws of thermodynamics. Hah! But since by convection the gases are rising up and cooling as they go, we know that a cold molecule cannot warm a hotter one. In fact it is my Warmist friends and adversaries who inadvertently prove that their own theories of man-made Global Warming and consequent Climate Change are false and hold no water whatsoever.

The Warmists make two errors. First of all they think that the atmosphere warms the Earth. They have it upside down. Ask a sand lizard! Secondly they conveniently forget the Laws of Thermodynamics. Energy always flows from the greater to the lesser, from hot to cold – never sideways on. Those glib statements about '…thus warming the lower atmosphere' are proven both wrong in practice and wrong in theory. There is not a single shred of empirical evidence to support the man-made Global Warming theory. Signed: Sand lizard!

Conclusion: It all depends…

Anthony Bright-Paul
Thursday, 31 July 2014

In bringing this book of essays and articles to a conclusion I am reminded of the famous wartime philosopher, Professor Joad of *The Brains Trust,* and his oft repeated opening gambit in debate – It all depends what you mean by…

As I read through both the articles of my sceptic friends and Professors and also my own essays, I see again and again that it all depends what you mean by… Much of the even angry debate between Alarmists and Sceptics is more often than not caused by a lack of definition. What is even more unhappy for me as a sceptic is to see the disagreements within the Skeptic ranks, for the simple reason that definitions have not been agreed upon. So we can see that much that is supposed to be science really belongs to the realms of philosophy, of logic and the correct use of language.

Let us take the simple word 'trap'. We can see that this simple word can arouse any amount of misunderstanding unless it is properly defined, particularly as regards heat. Ah! There is yet another word, which needs to be defined – 'heat'. In my article, 'Heat and a Thermos Flask' which was based on an exchange of emails with a fellow sceptic, we can see how easily misunderstandings can arise if the words 'trapping heat' are not clearly defined. Once we agree that heat is only trapped when 100 degrees in means also 100 degrees out, then we can all agree that heat is never trapped, but that temperatures can be increased in an enclosed space, albeit a car in the sun, an oven or a glasshouse, **but only while heat is being generated**. We all know that once the Sun sets or even a cloud passes, the heat in a car will dissipate by itself. The heat from an oven will likewise disperse once the oven is switched off.

The Alarmists are convinced that the Globe is warming and that this warming is caused by man and in particular by his adding an extra 2.9% to the trace gas Carbon Dioxide in the atmosphere by his burning of fossil fuels. What I hope is that in the course of this book the articles have all questioned what exactly is meant by warming. What is meant precisely by Global Warming? People get very hot under the collar and abusive without having clearly decided what they are talking about. Are we talking about the 'atmosphere'? Do we mean all four levels of the atmosphere to the TOA, top of the atmosphere? What precisely is meant by the expression Lower Atmosphere?

Again, can this warming be measured by a thermometer? Can one say that the Globe reached such and such a temperature on August 1st 2014, and a few days later the Global temperature was a tenth of a degree higher? The answer as we all know is that no such Global temperature is possible, for the very good reason as the Astrophysicist James Peden puts it – Where do you place the thermometer? Are we measuring the Earth's crust? The molten lava? The seas at depth? The ocean surface? Or are we measuring or attempting to read the temperature of the atmosphere, which is reckoned to be some 100kms from sea level to the TOA, top of the atmosphere, wherein there is a huge gradient of temperatures?

The Alarmists who aver that the Globe is warming even if it seems to any normal person to be following a perfectly understandable daily and seasonal warming and cooling, as night follows day, as winter follows summer, they insist that on average the Global temperature is rising. According to their reasoning this can only be ascribed to man's burning of these aforesaid fossil fuels. They argue that that Oxygen and Nitrogen being transparent lets the energy from the Sun through, but the Greenhouse gases, in particular Carbon Dioxide, capture or trap the energy that leaves a sun-warmed surface. That is in brief the Greenhouse Theory, GHT in short.

For them the Solar Constant is an article of faith. That means that the amount of energy that we receive from the Sun is constant, unvarying, or with so little variation as makes no difference. Yet for them the difference of the variation of 385 parts per million by volume to 400 parts per million by volume of CO2 has made huge difference! So we are to suppose that the TSI is constant. I refer my reader here to look again at the brilliant piece by Stephen Wilde, 'The Death Knell of Anthropological Global Warming', where the Alarmists resolutely refuse even to consider variations in the Sun's output, even though we all know of solar storms, of solar winds, of huge sunspots the size of this Earth and conversely sometimes the very lack of sunspots. To conclude that the TSI, total solar irradiance, is unchanging and makes no difference is simply to ignore well-known facts. To ascribe Global Warming to man and to ignore the mighty Sun is nothing short of lunacy.

Most Skeptics agree that the Earth has been warming since the end of the Little Ice Age, say 1850, but to ascribe this to man and to take no notice of the Sun's variations, which has caused all the previous Glacials and Inter-Glacials, is to inhabit a cloud-cuckoo land, divorced from reality.

The Canadian Geophysicist Norm Kalmanovitch wrote in his paper GAC MAC Temperature that there are essentially only two factors that control Global Temperature – Energy in and Energy out. Weather satellites launched in late 1978 have been measuring incoming and outgoing long wave radiation now for over 30 years. Any observed warming can then be accounted for by the increase in incoming solar energy. To suggest that Carbon Dioxide in an open system traps heat is plainly erroneous. Energy in must equal energy out. Ergo what plainly matters is the Energy In without the need to translate this into terms of temperature.

The most scientific paper in this book, parts of which might be difficult for the average layman, is 'Greenhouse Gases in the Atmosphere cool the Earth', yet the essence of this paper is about 'insulation'. After my own article 'Can air heat itself?' I received an email to the effect that I had denied that CO_2 was a Greenhouse Gas, and that these same GHGs insulate the world. So that brings up an important point. It all depends what you mean by the word 'insulation.'

Insulation is a two edged sword. We can easily see this in housing. In California where I sold homes in the San Fernando Valley the wood framed homes heated up very quickly and all needed air-conditioning to keep cool. On the other hand the older Spanish style Mexican homes with their thick walls, needed no air-conditioning. The thick walls helped to keep the houses cool. We can see the same difference with a double or a single glazed conservatory. The double-glazed one will warm up from the feeble winter sunshine much more slowly that a single glazed one, but on the other hand will retain its internal heat longer.

I have seen it argued thus, even by some Skeptics. 'Other things being equal' – whatever that means – 'the more Carbon Dioxide in the atmosphere the more warming will result.' Well, that is a profoundly illogical statement, because it can equally be argued the greater the insulation more cooling will result, which is precisely the argument that Robert Ashworth, Nasif Nahle and Hans Schreuder propose. And if the radiation from the Sun is greater than the radiation from the Earth then their argument must prove the more logical. Once again it is not a question of the science, but the interpretation of the known facts, and that is philosophy.

We have the same idiocy over the words 'climate change' which if doubted makes the Alarmists almost foam at the mouth with indignation. Of course, what the Alarmists infer is that the incidence of climatic events has been caused by man and by his wicked burning of the aforesaid fossil fuels. Just for the hell of it I looked up the Earthquakes that have occurred in the last seven days. Here are a few of them.

Magnitude-3.0 **Earthquake** Strikes in Santa Monica Bay
KTL- Jul 30, 2014
The **quake** was felt widely along the Santa Monica Bay coastline, but only as a "light" or "weak" shaking, according to the USGS ShakeMap. "We have just experienced what felt like a small **earthquake**," the Hawthorne Police Department tweeted. "No reports

A 4 **magnitude earthquake** shakes the Moroccan region of Nador
Morocco World News
 - 12 hours ago
Rabat – A 4 **magnitude** tremor was felt earlier today in the province of Nador, 317 miles north east of Rabat. The epicenter of the **earthquake**, which took place at 5:15 p.m. and lasted for 10 seconds, was located in the rural area of Kariat Arkmane, 16 ...

A **magnitude** 6.3 **Earthquake** hit the Eastern Mexican state of Veracruz
The Yucatan Times - Jul 29, 2014
A strong **magnitude** 6.3 **earthquake** shook much of eastern Mexico on Tuesday July 29th, 2014, a few minutes before 6 in the morning, but there were no reports of damage or injury. The U.S. Geological Survey said the**magnitude**-6.3 **quake** was centred in the ...

Earthquake of **magnitude** 5.7 reported off Papua New Guinea: USGS
Reuters - Jul 30, 2014
(Reuters) - An **earthquake** of **magnitude** 5.7 was recorded off the coast of Papua New Guinea on Wednesday, the United States Geological Survey reported. The USGS said the **earthquake's** epicenter was 133 kilometers south west of Arawa and 10 kilometers ...

It so happens that I lived in Santa Monica for a while and got married there moreover, so the first item was of great interest to me. As per usual the climate fanatics seem unable to distinguish between cause and effect. Are we to infer that these earthquakes are caused by Man?

Indonesia: Volcano nation

Indonesia is home to around 130 active volcanoes

As Indonesia faces yet another volcanic eruption, the BBC looks at some of the most recent volcanic blasts and at why Indonesia's islands are so volatile.

There are around 130 active volcanoes in Indonesia.

Another active volcano is Mt Sinabung on the Indonesian island of Sumatra, which burst into life in 2010, after 400 years of dormancy.

It became active again in September, after being dormant for three years.

Most recently, on 1 February, it launched a series of eruptions, spewing hot gas, ash and rocks 2km (1.5 miles) into the air and killing at least 14 people.

Mt Sinabung killed at least 14 people in a series of eruptions in February

I thought I might add some volcanoes to the Earthquakes. When you see these pictures of just two volcanoes in Indonesia, when you realise that there are hundreds of volcanoes not only above ground but also under the sea it makes you realise that those potentates who talk glibly about tackling climate change are simply massively ignorant or simply unwilling to face facts.

At the beginning of this year 2014, the Jetstream which blows way up in the Stratosphere altered its normal course southwards with the result there was a seemingly endless stream of depressions coming in from the Atlantic producing severe flooding in many parts of Somerset and the Thames Valley. Could this have been the result of climate change? Could this have been the fault of man burning fossil fuels? Or, perish the thought, was it the fault of Great Nature for producing 97.1% of this same Carbon Dioxide?

It all depends what you mean by 'tackling climate change?' What the Alarmists and the Alarmist politicians mean apparently is simply reducing the amount of Carbon Dioxide that man produces. Hey presto! This will at one and the same time head off Anthropogenic Global Warming and the insidious Climate Change that is concomitant. Can anyone take such childish ignorance seriously? One would think that nobody would be taken in by such obvious claptrap, yet that is precisely what has happened.

It is sometimes said even by Skeptic scientists that it is generally agreed that Carbon Dioxide warms the Lower Atmosphere. Here we have to revert to Professor Joad once again. What precisely is meant by the verb 'warm'? That seems a daft sort to question. How does a gas whose temperature varies with its surroundings, which is part of the mixture of gases which is called air, how is it supposed to warm anything? It is clear beyond peradventure that Carbon Dioxide does not generate heat – even the Alarmists agree to that. Then we know from the Thermos flask, which is the most efficient way we know of conserving heat <u>for a time</u>, but we know that after 24 hours 100°C of boiling water will have become tepid, so we know that heat cannot be trapped, then what is left?

The only thing that is left is back radiation. Well we are all agreed that every molecule above absolute zero radiates every which way according to its heat capacity. We also know that there is no correlation between temperature and the quantity of CO_2 in the atmosphere. That was demonstrated clearly by the graph in Professor Tim Ball's first email to me. The idea that a cold molecule could somehow radiate against the temperature gradient was clearly refuted by Dr Bratby in his exchange with Dr Helen Czerski. Apart from which from a purely philosophical and logical point of view back radiation does not work. If it did the radiation from the Earth back to the Sun would make the Sun hotter and the hotter Sun would make the Earth hotter and so on *ad infinitem.*

Such a scenario also fails from the point of view of the Inverse Square Law. The effectiveness of radiation diminishes with distance, which is why 'on the other side of our solar system, Pluto gets only a fraction of the heat,' as Hans Schreuder explains in 'Get your head round this.' When one considers the size of the Sun in relation to the Earth and the size of a molecule, which is too tiny to be visible, then it is clear that back radiation cannot work either.

This is put most succinctly by Dr Darko Butina in his paper 'Gas Laws and Greenhouse Theory or Back Radiation? What Back Radiation?' which I would have liked to include in its entirety within this book. However he writes: -

Conclusion: gas molecules of an open system are driven by temperature and it is physically impossible for gas molecules of the open system to control temperature in any shape for form.

http://www.principia-scientific.org/gas-laws-and-greenhouse-theory-or-back-radiation-what-back-radiation

So we can see that gas molecules are driven by temperature and do not drive temperature – and is that any news to any layman who listens to the Weather Forecasts every day? Carbon Dioxide does not control temperature in any shape or form. Absolutely. How can it possibly? How could it possibly? Any layman with a minimum of common sense and a modicum of reasoning power can see this. Carbon Dioxide is but a trace part of the mixture of gases that we call 'air', which the Rev Philip Foster answered to me in 'What's in the Space?'

Besides which, the atmosphere is an open system. What Back Radiation? asks Dr Darko Butina. 'Since the open system does not have any physical barrier it is impossible for that system to somehow 'radiate back' that extra heat!'

What have we done? What is the conclusion? If Carbon Dioxide does not generate heat, if Carbon Dioxide cannot trap heat and finally if Carbon Dioxide cannot radiate back heat, then all the business of reducing emissions of this CO2 are an exercise in utter fatuity.

I started out with the Great Global Warming Swindle and I am finishing with the Great Global Warming Hoax. Happily there are some who have not been taken in by this nonsense yet who are reviled and abused as Skeptics.

Why should those who question the so-called science of the Alarmists be subject to abuse? Why was the much-admired mathematician Johnny Ball cat-called? Why did the much-loved Biologist David Bellamy disappear from our TV screens, simply for declaring that green plants love CO2? Why did Tim Ball my first mentor suffer death threats? Why was Bob Carter treated so shabbily by his own University? Why has Lord Lawson, recognised by friend and foe for the brilliance of his intellect, why has he received totally uncalled for abuse? Even though he appears to accept to some degree the 'Greenhouse Effect' (which I do not) I have included his Bath lecture for its final conclusions and its final sentence, which is also now mine.

Global warming orthodoxy is not merely irrational. It is wicked.

Why is it wicked? Governments everywhere are throwing away our money on ridiculous schemes. Funding flows to Universities who look for a 'human fingerprint' whose remit is to prove that humans are causing Global Warming. Can you blame Universities for fighting for what seems a freebie, a free handout of funds from government? Can you blame a farmer or landowner if he can get a guaranteed income for say 25 years by covering good arable land with solar panels? So many landowners have made a packet by having Wind Turbines on their lands, all paid out of the public purse, that is taxation paid by you and me, to produce electrical power in the most inefficient way possible. So the scientists have advised governments and governments have created or fostered phoney industries. In the meantime more and more people become vassals, as more and more people benefit from grants, so the whole world economies are distorted, while the manufacturers of Wind Turbines and Solar Panels rub their hands in glee as governments approve and pass yet more and more useless inefficient schemes like Wind Farms covering acres and acres of sea and landscapes.

Even some sceptically inclined Members of Parliament boast that they understand the science. Just what does that mean? That Carbon Dioxide absorbs and emits infrared? But then so does every molecule. And so does a jar of pickles on a kitchen table, as Gary Novak points out. It absorbs and emits, but there is **no addition of heat**. Once again this is a matter of logic, not of science.

The juggernaut of greed and ignorance rolls on. No wonder that Nigel Lawson says: -

Global warming orthodoxy is not merely irrational. It is wicked.

Anthony Bright-Paul
02/08/2014

www.ingramcontent.com/pod-product-compliance
Lightning Source LLC
Chambersburg PA
CBHW071357170526
45165CB00001B/88